Understand Electrical and

Understand Electrical and Electronics Maths

Owen Bishop

Newnes
An imprint of Butterworth-Heinemann Ltd
Linacre House, Jordan Hill, Oxford OX2 8DP

 A member of the Reed Elsevier plc group

OXFORD   LONDON   BOSTON
MUNICH   NEW DELHI   SINGAPORE   SYDNEY
TOKYO   TORONTO   WELLINGTON

First published 1993
Reprinted 1994

© Owen Bishop 1993

All rights reserved. No part of this publication
may be reproduced in any material form (including
photocopying or storing in any medium by electronic
means and whether or not transiently or incidentally
to some other use of this publication) without the
written permission of the copyright holder except in
accordance with the provisions of the Copyright,
Designs and Patents Act 1988 or under the terms of a
licence issued by the Copyright Licensing Agency Ltd,
90 Tottenham Court Road, London, England W1P 9HE.
Applications for the copyright holder's written permission
to reproduce any part of this publication should be addressed
to the publishers

British Library Cataloguing in Publication Data
Bishop, O. N.
   Understand Electrical and Electronics Maths
   I. Title
   621.3810151

ISBN 0 7506 0924 9

Composition by Genesis Typesetting, Laser Quay, Rochester, Kent
Printed in Great Britain by The Bath Press, Avon

Contents

Preface vii
Symbols for mathematical operators ix
Symbols for equality and inequality xi

Part One – Basic Maths 1

1. About numbers 3
2. Quotients 12
3. Logs and other topics 23
4. Making sense of equations 33
5. Looking at graphs 48
6. From theory to practice 75

Part Two – Maths Topics 91

7. Distance and direction 93
8. Limits 103
9. Rates of change 111
10. Summing it up 138
11. Equations for actions 158
12. Practical differentials 180
13. Systematic solutions 212
14. Beyond the number line 240
15. Analysing waveforms 268

Answers 292
Index 313

Preface

Too often, the student who dips into a course textbook on electronics or electrical engineering is confronted by a line or two of unintelligible maths, and is thus prevented from going on to the more understandable electronic matters beyond. This book has been written to help the non-mathematical reader over such hurdles.

It is intended for students taking electronics and courses related to electrical engineering at levels up to and including BTEC Higher National Certificate and Diploma. It is also likely to be of use to first-year undergraduate students. Finally, it is hoped that the book will be interesting and instructive to those who follow electronics as a hobby and wish to extend their knowledge by private study.

The book is in two parts, intended as bridges to the understanding of maths as it applies to electrical and electronic circuits. Part One covers elementary maths, assuming a starting level a little below that of GCSE mathematics. It is a bridge between school maths and the basic maths required bor beginning a course on electronics. Part One is particularly intended to help the student who perhaps missed certain topics during early studies, or perhaps did not miss them but never quite understood them. It gives the student the chance to repair the gaps, in readiness for the special topics of the second part. But, as the last chapter of Part One shows, many aspects of electronics require surprisingly little specialist maths and much can be done with elementary maths.

At the beginning of each chapter in Part One the *Try these first* questions will indicate whether or not the student needs to spend time on the chapter. The *Test yourself* batches of questions occurring throughout the book are to assess progress at every stage.

Part Two deals with the more advanced maths topics required for electronics and electrical studies. It is a bridge between theoretical maths and its applications in practical circuits. The emphasis is on the maths, leaving the electronics to be covered by other books or by instructors. Only a minimum knowledge of electronic circuits is assumed. Once over this bridge, the student is ready to proceed to more advanced electronics without requiring much more in the way of mathematical expertise. The text concentrates on just those aspects applicable to electronics and electrical studies. For example, matrix maths is restricted to the use of matrices in solving simultaneous equations, ignoring the many other interesting aspects of matrices which have little relevance to electronics.

The main text is supplemented by material printed in boxes or panels. These are used to provide tables or brief summaries of procedures, intended to help the student grasp the essentials and remember them for subsequent use. Some are used to amplify the main discussion for those students who want to know more.

Although the maths in this book is aimed at practical problems based on practical circuits, there are certain topics which may be explored more widely and to advantage by the interested student. These topics are highlighted under the heading *Explore this*. Although these explorations are not mandatory, there is a lot to be gained by way of acquiring a feel for the subject in this way. An hour or so spent with a calculator or micro in exploring these byways will be more than repaid by the deepening of the student's understanding of electronics maths.

Symbols for mathematical operators

Operation	Symbol	Notes
Addition	$+$	
Subtraction	$-$	Connects *two* quantities, the second of which is to be subtracted from the other, for example, $a - b$.
Negation	$-$	Placed before a *single* quantity, this makes its value negative, for example $-a$ is the negative of a.
Positive or negative	$\pm$	The quantity may be positive or negative.
Multiplication	$\times$	Used if it makes an expression clearer.
	$\cdot$	Saves space and helps make expressions neater. Often we omit the multiplication sign. For example, ab means $a \times b$, $a(b + 2)$ means $a \times (b + 2)$.
	$*$	Used in some computer languages.
Division	$\div$	Traditional, but not often used now.
	$/$	Also used in some computer languages. The line may be horizontal for expressions with more than one term in them. For example $$\frac{a + 2}{b - 6}$$
	$^{-1}$	ab^{-1} means a divided by b (p. 24).
Square root	$\sqrt{}$	
Cube root	$\sqrt[3]{}$	
Absolute	$\lvert x \rvert$	Has the value of x, ignoring negation.
Increment	Δ	Δx is an increment (usually small) of x.
Function	$f(x)$ $g(x)$	$f(x) = x^2 - 3x + 4$. May also be written as f and g.
Differential coefficient	$\dfrac{\mathrm{d}x}{\mathrm{d}t}$	The differential of x with respect to t.
Partial differential coefficient	$\dfrac{\partial x}{\partial y}$	The differential of x with respect to y, other variables held constant.
Exponential	e^x $\exp x$	See pp. 5,9.
Natural logarithm	$\ln x$ $\log_e x$	
Common logarithm	$\lg x$ $\log_{10} x$	

$\text{Sin} = \dfrac{O}{H}$ $\text{Cos} = \dfrac{A}{H}$ $\text{Tan} = \dfrac{O}{A}$

S O H C A H T O A

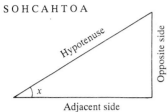

Complex operator	i, j	$i^2 + 1 = 0$. j preferred for electronics.
Laplace transform	$\mathcal{L}\{f(x)\}$	$\mathcal{L}\{e^{at}\} = 1/(s - a)$.
Sine	$\sin x$	Opposite side divided by hypotenuse.
Cosine	$\cos x$	Adjacent side divided by hypotenuse.
Tangent	$\tan x$	Opposite side divided by adjacent side.
Inverse sine	$\sin^{-1} x$ arcsin x	Similarly for cosine and tangent.

Symbols for equality and inequality

Symbol	Meaning	Notes
=	Equal	The expressions joined by this sign have the same value in the given calculation. In some computer languages, such as BASIC, this symbol assigns a value to a variable. For example, LET $A = A + 3$. In this, A does not *equal* $A + 3$.
≠	Not equal	
≈	Approximately equal	
≡	Is identical to	The expressions joined by this sign are necessarily equal owing to their mathematical properties. For example, $\tan \theta = \sin \theta / \cos \theta$.
>	Is greater than	
<	Is less than	
≥	Is equal to or greater than	
≤	Is equal to or less than	

Part One – Basic Maths

This part of the book is a summary of the essential maths that you need for the topics discussed in the second part. It is intended for revision and as a quick reference guide. There are a few practice exercises so that you can test your understanding of each section.

1 About numbers

Many people think of maths as a mass of numbers and symbols, which is true. This chapter tells you about numbers and some of the ways they are handled. Maths symbols are listed and explained in the table at the beginning of the book.

You need to know only a little elementary arithmetic and algebra to study this chapter. If you correctly answer all the questions below, you probably do not need to read this chapter.

Try these first

Your success with this short test will show you which parts of this chapter you already know.

1 From the numbers listed below, pick out those which are:

  a real numbers

  b rational numbers

  c natural numbers

  d integers

  1.2, 2, −6, ⅖, −⁵⁄₇, π, −0.7, 73, √5

2 Which of these are prime numbers? 7, 51, 2, 48, 1.9, 13

3 Factorize 286.

4 What is the HCF (highest common factor) of 60 and 78?

5 What is the LCM (lowest common multiple) of 24 and 30?

6 Factorize: **a** $4 - x^2$
  b $3ab^4 + 6a^2b^3$
  c $2x^2 - 9x + 10$

Types of numbers

Numbers have no units. They are not quantities of electric charge, or amounts of current. They are just numbers. They are represented on a **number line** (Figure 1.1), and are classified in various ways.

Natural numbers, or **counting numbers**, are the numbers we use when answering the question 'How many?'. They are always positive. For example, 1, 2, 3, 4, ... 100, ... and 2348, are all natural numbers.

Whole numbers are the natural numbers together with zero.

Integers comprise all the whole numbers and their negatives. For example, −14, −7, 0, 3, and 12 are all integers.

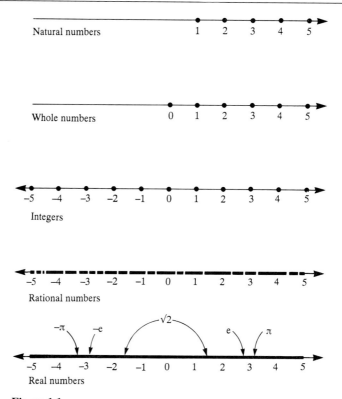

Figure 1.1

Rational numbers comprise all the integers and all numbers that can be represented as ratios. A ratio is one integer divided by another integer, provided that the dividing integer is not zero. Examples of ratios are 2/5, 34/67, and 233/8. The ratios fill up most of the length between the integers on the number line. Ratios can also be expressed as **decimal fractions**. For example, 2/5 is a ratio and is also written as 0.4. The decimal form of 2/5 terminates with one decimal place (1 dp). In other words, it needs only one decimal place to express 2/5 *exactly*. In some cases the ratio produces a **recurring decimal**. For example 7/3 is 2.333 333 333 . . . indefinitely. This is usually written 2.3̇.

Rational numbers include the natural numbers, whole numbers and integers because an integer such as 5, for example, can be represented by a ratio such as 20/4, or 5/1. Most values can be obtained by a suitable choice of numerator and divisor, so the rational numbers almost completely fill the number line. Figure 1.1 shows gaps among the rational numbers which indicate that there are some values that cannot be filled by ratios. The gaps are actually narrower than shown, and there are many more of them.

Real numbers comprise the rational numbers and the **irrational numbers**. Irrational numbers are numbers that are written as non-recurring, non-terminating decimals. The sequence of digits after the decimal point continues indefinitely without ever repeating. Examples of irrational numbers include

$\sqrt{2}$, π, e, 5.343 343 334 ..., and many others. These numbers fill up the remaining gaps in the number line.

> **Two important constants**
>
> π pi, the ratio of the circumference of a circle to its diameter, value = 3.1416 (to four decimal places).
>
> e the exponential constant, value = 2.7183 (to four decimal places). Useful for expressing rates of increase or decrease (page 59), and as the base of natural logarithms (page 26).

Imaginary numbers are explained on page 240. They do not appear on the number line of Figure 1.1. We use imaginary numbers in several branches of electronic theory, for example, when studying alternating signals.

Factors

An integer which divides exactly into another integer is called a **factor** of that integer. For example, 6 is a factor of 18, because it divides into 18 exactly 3 times, with no remainder. The complete list of factors of 18 is:

1, 2, 3, 6, 9, and 18

The list of factors of an integer always includes 1 and the integer itself. Some integers have only 1 and themselves as factors. Such integers are called **prime numbers**.

> **Prime numbers**
>
> The first twenty prime numbers are:
>
> 2, 3, 5, 7, 11, 13, 17, 19, 23, 29, 31, 37, 41, 43, 47, 53, 59, 61, 67 and 71.

We find the prime factors of an integer (or **factorize** it) by following the routine set out in the box.

Example
Factorize 84, listing (a) the prime factors and (b) all factors.
  (a) Finding the prime factors.

Step	Calculation	Factor recorded
Divide 84 by 2	84/2 = 42 exactly	2
Divide 42 by 2	42/2 = 21 exactly	2
Divide 21 by 2	21/2 does not go exactly	–
Divide 21 by 3	21/3 = 7 exactly	3
Divide 7 by 3	7/3 does not go exactly	–
Divide 7 by 5	7/5 does not go exactly	–
Next prime is 7		7

The prime factors of 84 are 2, 2, 3 and 7.

(b) Finding all factors.
1 is also a factor of 84. Other factors of 84 are obtained by multiplying together the prime factors in all possible combinations. In this example, we multiply them in twos, threes or fours:

$2 \times 2 = 4$
$2 \times 3 = 6$
$2 \times 7 = 14$
$3 \times 7 = 21$
$2 \times 2 \times 3 = 12$
$2 \times 2 \times 7 = 28$
$2 \times 3 \times 7 = 42$
$2 \times 2 \times 3 \times 7 = 84$

Multiplying all prime factors together gives the integer itself.

The factors of 84 are 1, 2, 2, 3, 4, 6, 12, 14, 21, 28, 42, and 84.

Finding factors

1 Divide the integer by 2.
2 Note if it divides exactly. If it does, repeat 1, using the quotient.
3 If division is inexact, begin again at 1 with the next highest prime (3, 5, 7 and so on).
4 Repeat 1 to 3 until the quotient is prime.

Highest common factor

The **highest common factor** of two or more integers is the highest number which is a factor of all of them. We refer to this as the **HCF**.

Quick guide to factors

2 is a factor if the number is even (ends in 0, 2, 4, 6 or 8).
3 is a factor if the sum of the digits is divisible by 3.
5 is a factor if the number ends in 5 or 0.
11 is a factor if the sums of the alternate sets of digits are equal.

Examples

In 2547, the sum of digits is 18; this is divisible by 3 and therefore 3 is a factor of 2547.
    In 2431, the sum of the 1st and 3rd digits is 5; the sum of the 2nd and 4th digits is 5; the sums are equal and therefore 11 is a factor of 2431.

Example
Find the HCF of 84 and 70.

The prime factors of 84 are 2, 2, 3, and 7.
The prime factors of 70 are 2, 5 and 7.

Set out the factors in columns.

   For 84:   2   2   3   7
   For 70:   2   5   7

The first 2 and the 7 are common to both integers.
Multiply these factors together: 2 × 7 = 14.

   The HCF of 84 and 70 is 14.

Finding the HCF

1 Find the prime factors of each number.
2 List them in columns, like below like.
3 Note which column or columns, if any, are *full*. These are the common factors.
4 Multiply them together to obtain the HCF.

The HCF of three or more numbers is found in a similar way.

Lowest common multiple

Given the prime factors of two or more integers, their **lowest common multiple** is the smallest number which has all these factors. We refer to this as the **LCM**.

Example
Find the LCM of 84 and 70.
   The first stages are exactly the same as above, giving

   for 84:   2   2   3   7
   for 70:   2   5   7

There are columns for 2, 2, 3, 5, and 7.
Multiplying these factors together: 2 × 2 × 3 × 5 × 7 = 420

   The LCM of 84 and 70 is 420.

The LCM of three or more numbers is found in a similar way.

Finding the LCM

1 Find the prime factors of each number.
2 List them in columns, like below like.
3 For each column, write down the value of the factor (the column does not have to be full).
4 Multiply the values together to obtain the LCM.

Test yourself 1.1

1 Which of these are integers? 3, 0.7, –5, 2/3, 0, –100
2 Which of these are rational numbers? 4/67, –56/11, 0.7, 0, 2π, 21
3 Which of these are real numbers? 12, –3/4, e, 0.625, 1.142856, 0
4 Which of these are natural numbers? 4/5, –7, 0.4, 56, 0, π

List all the factors of the following integers.

5 42   6 45
7 385   8 1950

Find the HCF of these sets of integers.

9 270, 770   10 2394, 1365
11 390, 273, 429   12 252, 330, 462

Find the LCM of these sets of integers.

13 330, 231   14 168, 420
15 16, 36, 90   16 126, 132, 165

Explore this

Under this heading you will find suggestions for topics to explore, to help you understand the maths involved. Use a calculator or microcomputer. Try writing your own programs in BASIC, C or PASCAL or make use of ready-made software. For later explorations you may need spread-sheets or graphics packages with graph-plotting features.

To start off, devise a routine to generate prime numbers up to 1000. If this works well, try extending it to factorize numbers and to find HCFs and LCMs.

Factors of polynomials

A **polynomial** is an algebraic expression in which there are two or more terms and in which powers of the x's, y's and other variables are all integers. Examples of polynomials are: $a^2 + 3b$, $x^3y - 5xy^2$, and $3p^2 + 4pq - 7q^2$.

The box lists methods for finding factors. Try the methods in order.

Factors of polynomials

1 Look for common factors; put them outside the bracket.
2 Look for common factors after terms have been regrouped.
3 Look for 'a square minus a square':
$a^2 - b^2 = (a + b)(a - b)$
4 Quadratic expression with three terms (see examples and next box).

Examples

Factorize $4ab + 6a$.
Try method 1, remembering to factorize the **coefficients** too. These are the numbers multiplying the algebraic constants. In this example the coefficients are 4 and 6.

Factors of $4ab$ are: 2   2 a b
Factors of $6a$ are: 2   3 a

The common factors (full columns) are 2 and a, so the HCF is $2 \times a = 2a$. Write the HCF in front of the brackets. Inside the bracket, for each term, write the products of those factors which do *not* belong to the HCF.

$\underline{4ab + 6a \text{ factorized is } 2a(2b + 3).}$

Factorize $6xy^2 - 9x^2y$.
Try method 1. The coefficients are 6 and 9.

Factors of $6xy^2$ are: 2   3 x y y
Factors of $9x^2y$ are: 3   3 x x y

The HCF is $3xy$.

$\underline{6xy^2 - 9x^2y \text{ factorized is } 3xy(2y - 3x).}$

Factorize $4a + 8 + ab + 2b$.
Method 1 shows that there are no common factors. With method 2, inspection shows that the expression factorizes if we group the first two terms and the last two terms.

Factors of $4a + 8$ are: 2   2 $(a + 2)$
Factors of $ab + 2b$ are: b $(a + 2)$

The HCF of the groups is $(a + 2)$

$$4a + 8 + ab + 2b = 4(a + 2) + b(a + 2)$$
$$= (4 + b)(a + 2)$$

$\underline{4a + 8 + ab + 2b \text{ factorized is } (4 + b)(a + 2).}$

Factorize $x^2 - 25$.
Methods 1 and 2 do not work, but method 3 does, for this is 'a square minus a square'.

$\underline{4x^2 - 25 \text{ factorized is } (2x + 5)(2x - 5).}$

Factorize $x^2 + 7x + 12$.
There are no common factors and it is not 'a square minus a square'. The highest power of x is x^2, so this is a ***quadratic expression*** with three terms. The signs are + and +.

Factorize the first term: $(x \quad)(x \quad)$

Factorize the last term so that the *sum* of the factors equals the coefficient of the middle term. Factors 1 and 12, or 2 and 6, do not add up to 7; but 3 and 4 do: $(x + 3)(x + 4)$

$\underline{x^2 + 7x + 12 \text{ factorized is } (x + 3)(x + 4).}$

Factorize $x^2 + 3x - 18$.
A quadratic expression with three terms. The signs are + and −.

Factorize the first term: $(x \quad)(x \quad)$
Look for a pair of factors whose *difference* equals the coefficient of the middle term. 1 and 18, or 2 and

9 are no good; but 3 and 6 differ by 3. Make the *bigger* factor positive and the *smaller* factor negative. $(x - 3)(x + 6)$

$x^2 + 3x - 18$ factorized is $(x - 3)(x + 6)$.

Factorize $x^2 - 3x - 10$.

A quadratic expression with three terms. The signs are – and –. Because the sign of the middle term is negative, make the *bigger* factor negative and the *smaller* factor positive.

$x^2 - 3x - 10$ factorized is $(x + 2)(x - 5)$.

Factorize $x^2 - 5x + 6$.

A quadratic expression with three terms. The signs are – and +. The last term is positive but the middle term is negative; both factors of the last term must be negated. Look for a pair of factors whose *sum* equals the coefficient of the middle term. The required factors are 2 and 3.

$x^2 - 5x + 6$ factorized is $(x - 2)(x - 3)$.

Factorize $4x^2 + 16x + 12$.

Method 1 shows that the terms have a common factor, 4. The first stage of factorizing gives $4(x^2 + 4x + 3)$. The expression in brackets is quadratic, with three terms, so there is a second stage of factorizing.

$4x^2 - 16x + 12$ factorized is $4(x + 1)(x + 3)$.

Factorize $3x^2 + 11x + 10$.

A quadratic expression with three terms. The coefficient of x^2 is greater than 1, but there are no common factors. This type of expression is factorized by trial and error. Experience helps.

Factorize the first term into $3x$ and x. $(3x \quad)(x \quad)$

Look for factors of the last term as in previous examples. The factors have to be summed to get the coefficient of the middle term, but now we need to sum one factor with **3 times** the other factor. Trying factors 2 and 5 $\Rightarrow 3 \times 2 + 5 = 11$. Put the factor which does *not* need to be multiplied in the same bracket as the $3x$. $(3x + 5)(x + 2)$

$3x^2 + 11x + 10$ factorized is $(3x + 5)(x + 2)$.

Test yourself 1.2

Factorize these expressions.

1. $5x^3 + x^2$
2. $2ab^2 + a^2b$
3. $a^2 - 9$
4. $4x + ax + 4y + ay$
5. $x^2 + 6x + 8$
6. $x^2 - 3x - 10$
7. $x^2 - 8x + 15$
8. $5p + p^2 + 5a + ap$
9. $m^2 - 5m$
10. $x^2 + 2x - 15$
11. $15x^2 - 13x - 20$
12. $6xy + 18y + 6x^2 + 18x$
13. $x^2 - x - 6$
14. $5x^2 + 25x + 20$
15. $3x^2 - 2x - 8$
16. $9x^2 - 16$
17. $2ax - 3by + bx - 6ay$
18. $36 + 33x - 3x^2$
19. $x^2 - 13x + 42$
20. $x^2 + 2x - 35$
21. $2x^2 + 13x + 15$
22. $9m^2 - 1$
23. $6x^2 + 17x + 12$
24. $8x^2 + 2x - 15$

As a revision exercise, work the 'Try these first' questions on page 3.

Summary of signs

Signs in expression	Look for ... of factors of last term	Signs in factors
+ +	Sum	(+) (+)
+ −	Difference	(+ bigger) (− smaller)
− −	Difference	(+ smaller) (+ bigger)
− +	Sum	(−) (−)

2 Quotients

The pages of an electronics book are typically sprinkled with quotients, ranging from simple ones such as

$$\frac{V}{R} \quad \text{and} \quad \frac{R_1 R_2}{R_1 + R_2}$$

to more complicated ones such as

$$\frac{Z_{\text{in}}}{1 - \dfrac{\beta A_o}{1 + \beta A_o}}$$

A quotient is just one value divided by another, but there are rules for handling quotients. Because quotients play such an important part in electronics, as well as in maths, it is important to be familiar with these rules. This section explains what do to.

You need to know about factors, HCFs and LCMs. All of these are explained in Chapter 1, pages 3–11.

Try these first

Your success with this short test will show you which parts of this chapter you already know.

1 Simplify: **a** $\tfrac{2}{3} + 1\tfrac{1}{5}$

   b $2\tfrac{1}{4} \times 2\tfrac{5}{9}$

   c $2\tfrac{3}{5} \div 1\tfrac{5}{7}$

2 Simplify: **a** $\dfrac{6a}{4a^2 + 12a}$

   b $\dfrac{2x}{7y} \times \dfrac{3y}{4z}$

   c $\dfrac{4}{x - 2} - \dfrac{2}{2x + 1}$

3 Express: $\dfrac{7x + 4}{x^2 - x - 6}$ as a partial fraction.

A quotient consists of two values or expressions, written above and below a horizontal line. The value or expression above the line, the **dividend**, is to be divided by the value or expression below the line, the **divisor**. For example, in the quotient $\frac{V}{R}$, the dividend is *V* and the divisor is *R*. Sometimes it is more convenient to draw a sloping line, as in *V/R*. If we substitute numbers for the symbols (in this example a number of volts divided by a number of ohms) and divide one number into the other, the result is not usually an integer. It generally consists of an integer plus a **fraction**. The fraction may itself be written as a quotient (or **vulgar fraction**) or as a **decimal fraction**. For example, if *V* = 12 and *R* = 5, then:

$$\frac{V}{R} = \frac{12}{5} = 2\tfrac{2}{5} \quad \text{or} \quad 2.4$$

The expression $\frac{12}{5}$, in which the dividend is greater than the divisor, is called an **improper fraction**, as it has not been divided out.

Handling fractions

When we are dealing with fractions, that is, quotients with numbers in them instead of symbols, we often use two different names for the dividend and divisor. The divisor is called the **denominator**. It tells us the *denomination* of the fraction, whether it is a number of halves, or quarters, or fifths, or twentieths and so on. This is the same idea as when currency is expressed in denominations such as pounds, pence, dollars or cents. The dividend is called the **numerator**. It tells us the *number* of halves, quarters, fifths . . . that the fraction represents.

To simplify the descriptions of the techniques, we shorten the term numerator to N and the term denominator to D. Thus, any fraction is of the form:

$$\frac{N}{D}$$

If the numerator and denominator have common factors, the fraction may be simplified by **cancelling** out the HCF.

Example

In $\frac{14}{35}$, the HCF of 14 and 35 is 7. Cancel the 7's by dividing both numerator and denominator by 7:

$$\frac{14}{35} = \frac{2}{5}$$

Rules of the arithmetic are stated in the boxes. Below we show examples of these rules in action.

Multiplying

$$\frac{2}{5} \times \frac{3}{7} = \frac{2 \times 3}{5 \times 7} = \frac{6}{35} \qquad \text{Multiply N's; multiply D's}$$

$$\frac{2}{5} \times \frac{5}{7} = \frac{2 \times 1}{1 \times 7} = \frac{2}{7}$$ Cancel 5's; then multiply N's and D's

---

Multiplying fractions

Turn mixed fractions into improper fractions:

$$\text{fraction} \times \text{fraction} = \frac{\text{product of N's}}{\text{product of D's}}$$

Cancel if possible.

With **mixed fractions** (integer and fraction):

$$2\tfrac{2}{5} \times 4\tfrac{1}{4} = \frac{12}{5} \times \frac{17}{4} = \frac{204}{20} = 10\tfrac{4}{20} = 10\tfrac{1}{5}$$

A fraction can be converted into any number of **equivalent fractions** by multiplying its N and its D by the same number (this is the opposite of cancelling):

$$\frac{1}{2} = \frac{2}{4} = \frac{3}{6} = \frac{4}{8} = \ldots$$

We make use of this when adding (see below).

Dividing

$$\frac{2}{3} \div \frac{8}{9} = \frac{2}{3} \times \frac{9}{8} = \frac{1}{1} \times \frac{3}{4} = \frac{3}{4}$$ Invert second fraction; cancel 2's, 3's; then multiply

---

Dividing fractions

1 Turn mixed fractions into improper fractions.
2 Turn the second fraction upside down.
3 Cancel if possible.
4 Multiply.

Adding and subtracting

We can add and subtract fractions only if they belong to the same denomination (*all* are halves, fifths, twentieths . . .). In other words, they must have the same D.

$$\frac{3}{11} + \frac{5}{11} = \frac{8}{11}$$ Adding fractions with the same D

> **Adding and subtracting fractions**
>
> Add integers of mixed fractions separately
>
> If D's are all the same:
> – add or subtract N's, with D written *once* below the line;
> – cancel if possible.
>
> If D's are not all the same:
> – find LCM of D's (i.e. their product, if they have no common factors);
> – find equivalent fractions with the LCM as their D's;
> – Add or subtract as above.

If they do not have the same D:

$$\frac{2}{5} + \frac{3}{8}$$ Required D is the LCM of 5 and 8 = 40

$$\frac{16}{40} + \frac{15}{40} = \frac{31}{40}$$ Convert to equivalent fractions, all with D = 40 (see 'Multiplying', above)

Another example with differing D's:

$$\frac{4}{5} + \frac{2}{7} - \frac{3}{4}$$ Required D is $5 \times 7 \times 4 = 140$

$$\frac{112}{140} + \frac{40}{140} - \frac{105}{140} = \frac{47}{140}$$ Convert to equivalent fractions with D = 140

Here, the fractions are mixed:

$$3\tfrac{5}{8} + 5\tfrac{4}{7} = 3 + 5 + \frac{5}{8} + \frac{4}{7} = 8 + \frac{35}{56} + \frac{32}{56} = 8 + \frac{67}{56} = 8 + 1\tfrac{11}{56} = 9\tfrac{11}{56}$$

> **Test yourself 2.1**
>
> Simplify.
>
> 1 $\dfrac{1}{3} + \dfrac{3}{4}$ 2 $\dfrac{4}{5} - \dfrac{1}{4}$
>
> 3 $1\tfrac{1}{2} + 2\tfrac{2}{5}$ 4 $5\tfrac{5}{8} - 4\tfrac{1}{2}$
>
> 5 $\dfrac{3}{4} \times \dfrac{5}{6}$ 6 $\dfrac{7}{8} \times \dfrac{3}{4}$
>
> 7 $\dfrac{3}{4} \div \dfrac{5}{8}$ 8 $3\tfrac{1}{3} \times \dfrac{2}{5}$
>
> 9 $3\tfrac{1}{4} \div 5\tfrac{1}{5}$ 10 $1\tfrac{1}{3} \div 3\tfrac{3}{5}$

Algebraic fractions

The rules for multiplying and other operations are the same as for ordinary numerical fractions. Look out for chances to cancel. For example, the fraction

$$\frac{x^2 - x - 6}{x^2 + 9x + 14}$$

can be simplified by factorizing the N and D, then cancelling the common factor:

Factorize $\Rightarrow \dfrac{(x + 2)(x - 3)}{(x + 2)(x + 7)}$

Cancel $(x + 2) \Rightarrow \dfrac{x - 3}{x + 7}$

Multiplying

$$\frac{x}{2} \times \frac{y}{3}$$

$$= \frac{xy}{6}$$

Dividing

$$\frac{4}{x + 1} \div \frac{2}{x - 2} = \frac{4}{x + 1} \times \frac{x - 2}{2}$$

$$= \frac{2(x - 2)}{x + 1} \qquad \text{Cancel 2's, then multiply}$$

Adding and subtracting

With identical D's:

$$\frac{2x}{5} + \frac{3y}{5}$$

$$= \frac{2x + 3y}{5}$$

With differing D's:

$$\frac{x}{2} + \frac{x}{3} \qquad \text{Required D is the product of 2 and 3, which is 6}$$

$$= \frac{3 \cdot x}{6} + \frac{2 \cdot x}{6}$$

The equivalent fraction of $\frac{x}{2}$ is $\frac{3x}{6}$,

The equivalent fraction of $\frac{x}{3}$ is $\frac{2x}{6}$.

In each case the N is multiped by LCM/D.

$$= \frac{5x}{6}$$

Here the D's have a common factor; find the LCM:

$$\frac{7x}{6} - \frac{4x}{15}$$

The LCM of 6 and 15 is 30

$$= \frac{5 \cdot 7x}{30} - \frac{2 \cdot 4x}{30}$$

Multiply each N by LCM/D

$$= \frac{35x - 8x}{30} = \frac{27x}{30}$$

Subtract; then we can cancel by 3

$$= \frac{9x}{10}$$

An example with an algebraic LCM:

$$\frac{4}{xy} + \frac{5}{x^2}$$

Required D is the LCM of xy and x^2, which is x^2y

$$= \frac{x \cdot 4}{x^2y} + \frac{y \cdot 5}{x^2y}$$

Multiply the N's by LCM/D. For the first fraction LCM/D = $x^2y/xy = x$

$$= \frac{4x + 5y}{x^2y}$$

Test yourself 2.2

Simplify.

1. $\dfrac{2x}{3x^2 + x}$ 2. $\dfrac{6x}{2x^2 + 4x}$

3. $\dfrac{x + x^2}{x + 1}$ 4. $\dfrac{5x + 15}{3x + 9}$

5. $\dfrac{x^2 + x - 6}{x^2 - 6x + 8}$ 6. $\dfrac{x^2 - 2x - 15}{x^2 - 9}$

Multiply.

7. $\dfrac{2x}{z} \times \dfrac{4y^2}{7}$

8. $\dfrac{3x}{x+2} \times \dfrac{2x-1}{x-2}$

Add or subtract.

9. $\dfrac{2x}{4} - \dfrac{x}{5}$

10. $\dfrac{3x}{4y} + \dfrac{2x+3}{4y}$

11. $\dfrac{2}{a} + \dfrac{3}{ab}$

12. $\dfrac{3x}{y^2z} - \dfrac{4x}{2y}$

13. $\dfrac{3}{x+1} + \dfrac{5}{x+3}$

14. $\dfrac{4}{x+1} - \dfrac{3}{x+2}$

15. $\dfrac{3}{x+1} + \dfrac{2}{x-1}$

16. $\dfrac{3}{(x+1)^2} - \dfrac{x}{x+1}$

Partial fractions

If it is difficult to perform an operation (such as differentiation, page 117) on a given fraction, it may help to split the fraction into two or more parts—its **partial fractions**. Finding partial fractions is the equivalent of adding fractions, but in reverse. For example,

$$\dfrac{2x}{x^2-1} = \dfrac{1}{x+1} + \dfrac{1}{x-1}$$

A fraction can be split into partial fractions only if the **degree** of the N (the highest power of x) is *lower* than the degree of the D. If this is not so, divide N by D (see final example below).

The D's of the partial fractions depend on the D of the original fraction. Factorize this first, if possible. Then write out the partial fractions it is to be broken into, using capital letters A, B, C and so on to represent the N's. The table in the box tells us what form the N's will take.

This example illustrates the stages of finding partial fractions.

Express $\dfrac{x}{(x+1)(x-2)}$ as partial fractions.

1 Preliminary inspection: The D is already in factors. It would contain x^2 if multiplied out, so the degree of the N is less than that of the D, as required to proceed.

2 Decide on form of partials: The D has two factors which differ. Both have the same form, $x+a$, so there will be two partial fractions of the form

$$\dfrac{A}{x+a}$$

Divisors of partial fractions

Factors in D	Partial fractions
$x + a$	$\dfrac{A}{x + a}$
$x^2 + ax + b$	$\dfrac{Ax + B}{x^2 + ax + b}$
$(x + a)^2$	$\dfrac{A}{x + a}$ and $\dfrac{B}{(x + a)^2}$
$(x + a)^3$	$\dfrac{A}{x + a}$, $\dfrac{B}{(x + a)^2}$ and $\dfrac{C}{(x + a)^3}$
$(x^2 + ax + b)^2$	$\dfrac{Ax + B}{x^2 + ax + b}$ and $\dfrac{Cx + D}{(x^2 + ax + b)^2}$

$$\Rightarrow \frac{x}{(x + 1)(x - 2)} \equiv \frac{A}{x + 1} + \frac{B}{x - 2}$$

This is an **identity**, an expression which is true for all values of x.

3 Clear fractions: Multiply throughout by the D of the original.

$$\Rightarrow x = A(x - 2) + B(x + 1)$$
$$\Rightarrow x = Ax - 2A + Bx + B$$

4 Equate the coefficients: The coefficients of terms with the same powers of x can be put equal to each other. The x on the left side has the coefficient 1. The x's on the left side have the coefficients A and B. We can write:

$$1 = A + B \tag{1}$$

Similarly, the constant terms (terms without x) have the coefficient 0 on the left side because there are no such terms. On the right, the coefficients are $-2A$ and B:

$$0 = -2A + B \tag{2}$$

5 Solve equations (1) and (2): as simultaneous equations (page 41).

$$(1) - (2) \Rightarrow 1 = 3A$$
$$\Rightarrow A = 1/3$$

Substitute in (1) $\Rightarrow B = 2/3$

6 Write out the partial fractions, using the values of A and B obtained at step 5:

$$\frac{x}{(x + 1)(x - 2)} = \frac{1}{3(x + 1)} + \frac{2}{3(x - 2)}$$

Another example:

Express $\dfrac{x^2}{(x + 1)^3}$ in partial fractions.

1 Preliminary inspection: The D is already in factors. It would contain x^3 if multiplied out, so the degree of the N is less than that of the D, as required to proceed.
2 Decide on form of partials: The D has three *identical* factors of form $x + a$, so there are three partial fractions of the form

$$\frac{A}{(x+a)}, \quad \frac{B}{(x+a)^2}, \quad \text{and} \quad \frac{C}{(x+a)^3}$$

$$\frac{x^2}{(x+1)^3} = \frac{A}{(x+1)} + \frac{B}{(x+1)^2} + \frac{C}{(x+1)^3}$$

3 Clear fractions:

$$\Rightarrow x^2 = A(x+1)^2 + B(x+1) + C$$
$$\Rightarrow \quad = A(x^2 + 2x + 1) + B(x+1) + C$$
$$\Rightarrow \quad = Ax^2 + (2A+B)x + A + B + C$$

4 Equate the coefficients:

$$\begin{aligned}
\text{Coefficients of } x^2 \quad & 1 = A & (1)\\
\text{of } x \quad & 0 = 2A + B & (2)\\
\text{of constants} \quad & 0 = A + B + C & (3)
\end{aligned}$$

5 Solve equations (1) to (3):

(1) $\Rightarrow A = 1$
Substitute in (2) $\Rightarrow B = -2$
Substitute in (3) $\Rightarrow C = 1$

6 Write out the partial fractions:

$$\frac{x^2}{(x+1)^3} = \frac{1}{(x+1)} - \frac{2}{(x+1)^2} + \frac{1}{(x+1)^3}$$

A mixed example:

The denominator has factors of two *different forms* so the partial fractions are taken from two lines of the table.

Express $\dfrac{x^2 + 10x + 10}{(x+2)(x^2 + 3x - 2)}$ as partial fractions.

1 Preliminary inspection: the quadratic expressions in N and D do not factorize, so the fraction cannot be simplified by cancelling. The order of the N is less than the degree of the D.
2 Decide on form of partials: The D has terms of the form $(x + a)$ and $(x^2 + ax + b)$, so the partial fractions are those shown in the first two lines of the table.

$$\frac{5x^2 + 18x + 4}{(x+2)(x^2 + 3x - 2)} = \frac{A}{x+2} + \frac{Bx + C}{x^2 + 3x - 2}$$

3 Clear fractions:

$$\Rightarrow 5x^2 + 18x + 4 = A(x^2 + 3x - 2) + (Bx + C)(x + 2)$$
$$= Ax^2 + 3Ax - 2A + Bx^2 + (2B + C)x + 2C$$

4 Equate coefficients:

$$5 = A + B \quad (1)$$
$$18 = 3A + 2B + C \quad (2)$$
$$4 = -2A + 2C \quad (3)$$

5 Solve equations (1) to (3)

Change subject of (1) $\Rightarrow B = 5 - A \quad (4)$

Divide (3) by 2 $\Rightarrow 2 = -A + C \quad (5)$

Substitute (4) in (2) $\Rightarrow 18 = 3A + 2(5 - A) + C$
$$= A + 10 + C$$
$$\Rightarrow 8 = A + C \quad (6)$$

Add (5) to (6) $\Rightarrow 10 = 2C$
$$\Rightarrow C = 5 \quad (7)$$

Substitute (7) in (3) $\Rightarrow A = 3 \quad (8)$

Substitute (8) in (1) $\Rightarrow B = 2$

6 Write out the partial fractions:

$$\frac{x^2 + 18x + 4}{(x + 2)(x^2 + 3x - 2)} = \frac{3}{x + 2} + \frac{2x + 5}{x^2 + 3x - 2}$$

N of higher degree than D:

In this example the N is of degree 3 but the D is only of degree 2:

$$\frac{x^3 + x^2 + x + 1}{x^2 + x - 6}$$

Divide N by D $\Rightarrow x + \dfrac{7x + 1}{x^2 + x - 6}$

Now find partial fractions of the quotient on the right, after factorizing the D:

$$\Rightarrow x + \frac{3}{x - 2} + \frac{4}{x + 3}$$

The result is the x from the first division, plus the partial fractions.

Test yourself 2.3

Express as partial fractions.

1. $\dfrac{5x - 23}{(x - 3)(x - 7)}$

2. $\dfrac{4x + 21}{(x - 1)(x + 4)}$

3. $\dfrac{3x + 5}{x^2 - x - 12}$

4. $\dfrac{x^2 + 1}{(x + 2)^3}$

As a revision exercise, work the 'Try these first' questions on page 12.

3 Logs and other topics

> This chapter covers indices, logs and vectors. Indices and logs are important in a wide range of calculations, while vectors are particularly useful in connection with the theory of alternating currents and voltages.
>
> You need only a little elementary maths for this chapter.

> **Try these first**
>
> Your success with this short test will show you which parts of this chapter you already know.
>
> 1 Simplify: **a** $x^4 \times x^3$
>   **b** $(a^4)^3$
>   **c** $\sqrt{b^x}$
>
> 2 If $e^{2.5} = 12.18$, what is the value of $\ln 12.18^2$?
>
> 3 Solve for x: $\log_x 27 = 3$
>
> 4 **a** Calculate the magnitude and direction of these vectors:
>
> $$\mathbf{A} = \begin{pmatrix} 4 \\ 3 \end{pmatrix} \qquad \mathbf{B} = \begin{pmatrix} -2 \\ 7 \end{pmatrix}$$
>
>   **b** State the value of their sum **C** as a matrix and calculate its magnitude and direction.

Indices

The power of a number is indicated by an **index**, sometimes called an **exponent**. For example, the index '5' in a^5 means five a's multiplied together. If the index is '1', this means just the number itself ($a^1 = a$).

In algebra, only terms of the same **order** (meaning, having the same index) can be added or subtracted:

$$2a^3 + 5a^3 = 7a^3$$

Terms of different order cannot be added or subtracted. For example, $3a^3 + 4a^2$ cannot be added to make a single term. But terms of different (or the same) order can be multiplied and divided. The box gives the rules for multiplication and division of indexed terms.

The box shows that $a^{-n} = 1/a^n$. These are two ways of writing out the same thing. One special case of this is when $n = 1$:

$$a^{-1} = \frac{1}{a}$$

The quantity $1/a$, is called the **reciprocal** of a, or the **inverse** of a.

Index rules

Rule	Example
$a^m \times a^n = a^{m+n}$	$a^2 \times a^5 = a^7$
$\dfrac{a^m}{a^n} = a^{m-n}$	$\dfrac{a^5}{a^2} = a^3$
$(a^m)^n = a^{mn}$	$(a^2)^5 = a^{10}$
$a^{-n} = \dfrac{1}{a^n}$	$5^{-2} = \dfrac{1}{5^2} = \dfrac{1}{25}$
$a^0 = 1$	$3^0 = 1$
$a^{1/2} = \sqrt{a}$	$9^{1/2} = \sqrt{9} = 3$
$a^{1/3} = \sqrt[3]{a}$	$8^{1/3} = \sqrt[3]{8} = 2$

The negative sign can also be used with other powers. For example:

a^2 means the square of a

a^{-2} means the reciprocal of the square of a ($= 1/a^2$)

The symbol $^{-1}$ is also used to indicate an operation which is the *inverse* of a given operation.

For example:

Taking the square root is the *inverse* of squaring. Starting with a we square it and get a^2. Then, given a^2, we take its square root (its inverse square) and get back to a.

Another example:

$\sin \theta = x$ means the sine of the angle θ is x

$\sin^{-1} x = \theta$ means the angle which has sine equal to x is θ.

Starting with θ, we take its sine and obtain x. Then, given x, we take its *inverse sine* and get back to θ again.

This **inverse notation** is also used with logarithms and vectors (see below) and also with matrices (page 225).

The bottom two lines of the box explain the meaning of **fractional powers**, that is, powers for which the index is not an integer. Other fractional powers are built up according to the same rules. For example:

$a^{2/3} = (a^{1/3})^2 = (\sqrt[3]{a})^2$

Similarly:

$a^{0.7} = a^{7/10} = (a^{1/10})^7 = (\sqrt[10]{a})^7$

It is easy to evaluate fractional powers by using logarithms (below), or on a calculator that has an x^y key.

Test yourself 3.1

Simplify.

1. $a^4 \times a^2$
2. $\dfrac{15b^4}{3b^4}$
3. $(a^3)^2$
4. $(3a^3)^2$
5. $4p^3 \times 2p^5$
6. $\dfrac{(3x^2)^4}{9x^2}$
7. $a^{-2} \times a^4$
8. $n^3 \times n^{-1} \times n^{-2}$
9. $\sqrt{a^4}$
10. $\sqrt[3]{27b^6}$

Logarithms

When we work with logarithms we are working with indices. For example, we know that $2^4 = 16$. Instead of using the number 16, we use the index 4, and call it the logarithm (or 'log') of 16. Provided that we know that we are dealing with powers of 2 (the **base number**) we can always turn numbers into logs and logs into ordinary numbers. Putting this in the form of an equation:

$$\log_2 16 = 4$$

Note the base number written as a subscript to the symbol 'log'.

Although *any* positive number except zero can be used as a base for logs, we generally use only 10 or e (= 2.7183, see page 5).

Logs to base 10

The relationships are expressed in this pair of equations:

$$a = 10^b \Rightarrow \log_{10} a = b$$

For example,

$$10 = 10^1 \Rightarrow \log_{10} 10 = 1$$

$$10\,000 = 10_4 \Rightarrow \log_{10} 10\,000 = 4$$

The equations apply to fractional indices too; for example,

$$47 = 10^{1.67} \Rightarrow \log_{10} 47 = 1.67 \text{ (2 dp)}$$

Negative indices occur for logs of numbers less than 1; for example,

$$0.01 = 10^{-2} \Rightarrow \log_{10} 0.01 = -2$$

$$0.73 = 10^{-0.14} \Rightarrow \log_{10} 0.73 = -0.14 \text{ (2 dp)}$$

Note that the log of zero and numbers less than zero are undefined. Logs to base 10 are so widely used that the symbol is often written without specifying the base number. We just write 'log' or 'lg'.

Logs to base e

These are known as **natural logarithms**. The equations are:

$$a = e^b \Rightarrow \log_e a = b$$

For example,

$$47 = e^{3.85} \Rightarrow \log_e 47 = 3.85$$

ln is an alternative symbol for $\log_e$.

Logs, being powers of the base number, are treated in the same way as indices. The box lists the rules.

Log rules

Rule	Example
$\log (m \times n) = \log m + \log n$	$\log (4 \times 5) = \log 4 + \log 5$
$\log (m/n) = \log m - \log n$	$\log (4/5) = \log 4 - \log 5$
$\log m^n = n \log m$	$\log 4^5 = 5 \log 4$
$\log (1/m) = -\log m$	$\log (1/4) = -\log 4$
$\log 1 = 0$	
$\log (\sqrt{m}) = \dfrac{\log m}{2}$	$\log (\sqrt{7}) = \dfrac{\log 7}{2}$

Before the days of pocket calculators, logs were very commonly used as a calculating aid. Adding or subtracting logs is much easier than performing long multiplication and division. The slide rule had logarithmic scales to make addition and subtraction of logs even quicker. Although logs are no longer used for that purpose, they often appear in equations and formulas so we still need to know the rules for handling them. These are summarized in the box.

Summary of index and log rules

To multiply   – ADD
To divide   – SUBTRACT
To square   – DOUBLE
To square-root – HALVE
To invert   – NEGATE

Logarithms are useful for solving equations containing x as an index.

Example
Solve $3^x = 7$. Take logs of both sides of the equation

$$\log (3^x) = \log 7$$
$$\Rightarrow \quad x \log 3 = \log 7$$

> **Rounding**
>
> Draw a line through the number to the right of the last place required.
>
> If the number to the right of that line is 4 or less, leave the number to the left of the line unchanged.
>
> If the number to the right of that line is 5 or more, increase the number to the left of the line by 1.
>
> | *Examples* | *Number to right is* | *Result* |
> |---|---|---|
> | Round 3.443 to 2 dp: 3.44\|3 | 4 or less | 3.44 |
> | Round 5.37 to 1 dp: 5.3\|7 | 5 or more | 5.4 |
> | Round 6.3747 to 2 dp: 6.37\|47 | 4 or less | 6.37 |
>
> *Note*: Round in *one* step: do not round to 6.735, then to 6.38.
>
> | | | |
> |---|---|---|
> | Round 7.995 to 2 dp: 7.99\|5 | 5 or more | 8.00 |
>
> *Note*: Rounding may carry forward to previous figures.

$$\Rightarrow \quad x = \frac{\log 7}{\log 3}$$

Evaluate, using a calculator $\quad x = \dfrac{0.8451}{0.4771} = 1.771$ (3 dp)

The same *result* is obtained if we use logs to base e instead, though the values of the logs used are different.

Logs are useful for evaluating fractional powers. In the following example we use logs to base 10, obtained from a calculator.
Evaluate $\quad x = 3.5^{0.61}$

$$\Rightarrow \log x = 0.61 \times \log 3.5 = 0.61 \times 0.5411 = 0.3319 \text{ (4 dp)}$$

Having obtained log x, we convert it to ordinary numbers by the *inverse* operation $\log^{-1}$, sometimes referred to as taking the **antilogarithm**. We use a calculator to find:

$$\Rightarrow x = \log^{-1} 0.3319 = 2.1473 \text{ (4 dp)}$$

The expression '$\log^{-1} 0.3319$' can also be written 'antilog 0.3319'.

One property of logs is the way that they compress the range of values of ordinary numbers. A range from 1 to 1 million may be compressed to a range from 0 to 6 by taking logs to base 10. This is useful when representing data graphically. The degree of compression varies along the scale. The smaller values (say, 1 to 100) are spread over the range 0 to 2, while the larger values (say, 10 000 to 1 000 000) are more compacted over the range 4 to 6. This makes it possible to show wide-ranging data on a single diagram. Plotting on a log scale also brings other benefits, as explained on page 61.

Test yourself 3.2

1. Evaluate.
   - **a** lg 100
   - **b** lg 1000
   - **c** lg 1 000 000

2. For each of the expressions below, write the equivalent in logarithmic form.
   - **a** $2^3 = 8$
   - **b** $5^4 = 625$
   - **c** $3^0 = 1$
   - **d** $36^{1/2} = 6$
   - **e** $5^{-3} = 0.008$
   - **f** $25^{-1/2} = 0.2$

3. For each of the expressions below, write the equivalent in index form.
   - **a** $\log_2 64 = 6$
   - **b** $\log_3 1 = 0$
   - **c** $\log_5 5 = 1$
   - **d** $\log_4 64 = 3$

4. Simplify.
   - **a** $\log 4 + \log 2$
   - **b** $\log 10 - \log 2$
   - **c** $2 \log 3$
   - **d** $\tfrac{1}{2} \log 16$
   - **e** $\log 3 + 2 \log 7$
   - **f** $3 \log 4 + 2 \log 5 - \log 20$

5. Solve for x.
   - **a** $\log_x 9 = 2$
   - **b** $\log_2 8 = x$
   - **c** $4^x = 5$
   - **d** $0.25^x = 9$

Vectors

Many quantities, such as distance, charge, potential difference (p.d.), and resistance, are completely expressed by a single number, usually with a unit. These are known as **scalar** quantities. The number tells us the size or magnitude of the quantity. Other quantities, such as force and electric field need *two* numbers to express them; one number (with a unit) for their *size* and a second number (also with a unit) for their *direction*. Quantities which have both size and direction are called **vector** quantities.

We indicate that a quantity is a vector by printing its symbol in bold type: **A**. In some books it is indicated by a caret (⌢) above the symbol: $\widehat{A}$. In handwriting, vectors have a wavy line under the letter: A̰.

A vector may be represented by:

(a) **Drawing:** as in Figure 3.1. The length of the line represents the magnitude (drawn to scale); the arrow shows the direction, from the *starting point* to the *finishing point*. The direction is measured anticlockwise from the positive x-axis.

(b) **Matrix:** containing its horizontal and vertical components. The matrices for the vectors shown in Figure 3.1 are:

$$\mathbf{A} = \begin{pmatrix} 6 \\ 3 \end{pmatrix} \qquad \mathbf{B} = \begin{pmatrix} -4 \\ -5 \end{pmatrix}$$

Figure 3.1

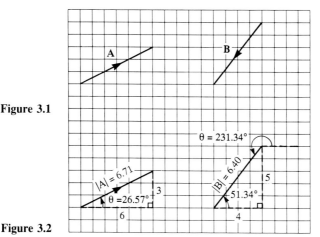

Figure 3.2

The horizontal component is written above the vertical component. Matrices do not tell us the direction or magnitude directly, but these can be calculated, as in Figure 3.2. By Pythagoras' theorem, the magnitude of vector **A** above is:

$$|\mathbf{A}| = \sqrt{(6^2 + 3^2)} = 6.71 \text{ (2 dp)}$$

We use vertical lines on either side of the symbol to indicate magnitude. The direction of **A** is:

$$\theta = \tan^{-1} 3/6 = 26.57° \text{ (2 dp)}$$

Note the use of the *inverse tangent* symbol (see page 24).

For vector **B**, the angle in the triangle obtained by calculating $\tan^{-1} 5/4$, is 51.34°, but the direction of the vector with respect to the positive *x*-axis is 51.34° + 180° = 231.34° (2 dp).

Angles of vectors can also be expressed in radians (see box).

Measuring angles

Angles are measured in degrees (symbol, °) or radians (symbol, rad).

180° = π rad

Converting degrees to radians: multiply by $\dfrac{\pi}{180}$ = 0.01745 (4 sf).

Converting radians to degrees: multiply by $\dfrac{180}{\pi}$ = 57.30 (4 sf).

Two vectors are equal if they have the same magnitude and direction. One vector is the inverse or negative of another vector if it has the same magnitude but the opposite direction. The symbol for the inverse of **A** is $\mathbf{A^{-1}}$.

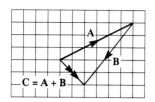

Figure 3.3

Vector addition

Vectors may be added by adding the matrices. We simply add the corresponding numbers. Given vectors **A** and **B** above:

$$\mathbf{A} + \mathbf{B} = \begin{pmatrix} 6 \\ 3 \end{pmatrix} + \begin{pmatrix} -4 \\ -5 \end{pmatrix} + \begin{pmatrix} 2 \\ -2 \end{pmatrix}$$

Another way of adding vectors is by drawing them end to end (Figure 3.3), the starting point of the second vector (**B**) coinciding with the finishing point of the first vector (**A**). Their sum or **resultant** vector (**C**) runs from the starting point of the first vector to the finishing point of the second vector. The resultant vector is marked with a double arrow. To add vectors in this way we either make a scale drawing and measure the magnitude and direction of the resultant from this, or we draw a sketch and use Pythagoras' theorem and trigonometry to calculate the dimensions and angles required. The figure shows the result of the addition, giving the resultant matrix:

$$\mathbf{C} = \begin{pmatrix} 2 \\ -2 \end{pmatrix}$$

A particular case of adding vectors occurs when they are perpendicular (Figure 3.4). **A** and **B** act at point O. Draw **B'** with its starting point coincident with the finishing point of **A**. **B'** is equal to **B** since it has equal magnitude and direction. Their sum is the diagonal OP. We use the magnitudes of **A** and **B** in the calculations.

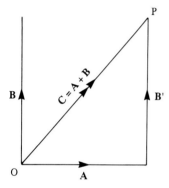

Figure 3.4

By Pythagoras' theorem:

$$|\mathbf{C}| = |\mathbf{A} + \mathbf{B'}| = \sqrt{|\mathbf{A}|^2 + |\mathbf{B}|^2}$$

By trigonometry:

$$\theta = \tan^{-1} \frac{|\mathbf{B}|}{|\mathbf{A}|}$$

Example
Given $|\mathbf{A}| = 4$, $|\mathbf{B}| = 7$

$$|\mathbf{C}| = \sqrt{4^2 + 7^2} = 8.06 \text{ (2 dp)}$$

$$\theta = \tan^{-1} 7/4 = 60.26° \text{ (2 dp)}$$

The reverse operation, resolving a vector into two **components** in perpendicular directions is as follows. Given θ, the components of **A** are **B** and **C** (Figure 3.5):

$$|\mathbf{B}| = |\mathbf{A}| \sin \theta$$
$$|\mathbf{C}| = |\mathbf{A}| \cos \theta$$

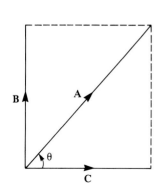

Figure 3.5

Usually, θ is the angle between the vector **A** and the x-axis.

Example

Given $|\mathbf{A}| = 5$, $\theta = 60°$

$|\mathbf{B}| = 5 \sin 60° = 5 \times 0.8660 = 4.33$ (2 dp)

$|\mathbf{C}| = 5 \cos 60° = 5 \times 0.5000 = 2.50$ (2 dp)

Test yourself 3.3

Draw these vectors.

1 $\mathbf{A} = \begin{pmatrix} 4 \\ 2 \end{pmatrix}$ **2** $\mathbf{B} = \begin{pmatrix} -3 \\ 5 \end{pmatrix}$ **3** $\mathbf{C} = \begin{pmatrix} 0 \\ -2 \end{pmatrix}$

Express these vectors as matrices.

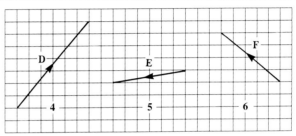

Figure 3.6

Calculate the magnitude and direction of these vectors (to two decimal places).

7 $\mathbf{A} = \begin{pmatrix} 3 \\ -3 \end{pmatrix}$ **8** $\mathbf{B} = \begin{pmatrix} 4 \\ 1 \end{pmatrix}$ **9** $\mathbf{C} = \begin{pmatrix} -5 \\ 6 \end{pmatrix}$

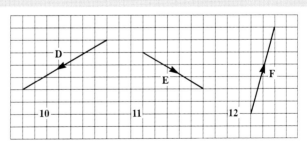

Figure 3.7

Given that

$\mathbf{A} = \begin{pmatrix} 3 \\ 4 \end{pmatrix}$, $\mathbf{B} = \begin{pmatrix} -1 \\ 2 \end{pmatrix}$, $\mathbf{C} = \begin{pmatrix} 2 \\ 2 \end{pmatrix}$, and $\mathbf{D} = \begin{pmatrix} -3 \\ -2 \end{pmatrix}$

in questions 13 and 18 calculate the matrix of the resultant vector. Then find to two decimal places by drawing or calculation, the magnitude and direction of the resultant vector.

Hint for questions 14 and 18: subtract the vector by finding the inverse vector and *adding* it.

13 A + B   **14** A − B   **15** A + C
16 B + D   **17** A + C + D   **18** B + D − C

Find the components of these vectors, given their magnitude and direction.

19 $|A| = 7$, $\theta = 20°$ **20** $|B| = 4$, $\theta = 45°$
21 $|C| = 3$, $\theta = 130°$

As a revision exercise, work the 'Try these first' questions on page 23.

4 Making sense of equations

Ohm's Law states that:

> The current passing through a conductor at constant temperature is proportional to the potential difference between its ends.

Expressed as an equation, Ohm's Law becomes:

$$I = \frac{V}{R}$$

The equation is the more convenient form of the Law because it clearly shows the relationship between current, p.d. and resistance. In particular, it tells us how to calculate any one of these quantities, given the other two. Equations are a shorthand way of stating physical relationships. An equation makes it easier to understand the relationships and to deduce other, perhaps more complicated, relationships. This is why equations are so important in electronics.

You need to know about factorizing, quotients, power and logs, all of which are discussed in the first three chapters of this book.

Try these first

1 Change the subject of this equation to R_B.

$$V = \frac{V_{BE}(R_A + R_B)}{R_B}$$

2 Solve these equations for x (answer to two decimal places).

   a $x^2 - 3x - 18 = 0$ **b** $2x^2 + 7x - 6 = 0$

3 Solve these simultaneous equations for x and y.

   a $3x + 5y = 1$ **b** $-2x + y = 11$
      $2x - 3y = 26$ $xy - 3 = -18$

4 Solve these simultaneous equations for x, y and z.

   $2x + y - 3z = -1$
   $3x - y + 6z = 5$
    $x + 2y - 5z = 4$

5 **a** Round 74 900 to two significant figures.
   b Round 25.749 to four significant figures.

6 **a** Round 5.369 51 to three decimal places.
   b Round 1.320 99 to two decimal places.

> **7** Express in scientific form **a** 2473
> **b** 0.052
>
> **8** Express in engineering form **a** 32.6×10^2
> **b** 537.42×10^{-8}

Types of equations

An equation states that two expressions are equal in value. An **identical equation** or **identity** states that the two expressions are equal for *all* values of the variables. Examples of identities are:

$$x^2 - y^2 = (x - y)(x + y)$$

$$\tan \theta = \frac{\sin \theta}{\cos \theta}$$

We often indicate that an equation is an identity by using the symbol ≡ instead of =.

Most of the equations met with in this book are **conditional equations**. The two sides of the equation are equal only for a few values (perhaps only for one value) of the variables. An example of a conditional equation is:

$$x^2 + 3x = 10$$

Given such an equation we usually try to solve it to find the values of x for which this equation is true. In this example, the two sides of the equation are equal only if $x = -5$ or $x = 2$.

Balancing equations

We often need to simplify, expand or otherwise alter one or both sides of an equation, in order to get it into the right form for some subsequent operation. The equation begins by being *balanced*; both sides are equal in value. If they are not, it is not an equation. When handling an equation, it is essential to *keep it balanced*. If this is not done, it ceases to be an equation. Below we list some of the more useful operations that can be performed on an equation while keeping it balanced at the same time. The vital rule is:

Always do the SAME thing to BOTH sides

The changes that are allowed include the following:

1 Rearrange terms, keeping them on the same sides, or carry out any implied operation such as removing brackets, or any other operations that do not affect the value of the side.

$4 + x^2 - 2x = -y + 2$ can become $x^2 - 2x + 4 = 2 - y$

$x = x(2 + y)$ can become $x = 2x + xy$

2 Exchange sides.

$3x + 6 = 4ab^2$ becomes $4ab^2 = 3x + 6$

3 Negate both sides.

$5p - q = -3$ becomes $-5p + q = 3$

This looks better if rearranged $\Rightarrow q - 5p = 3$

4 Add the same term to both sides.

$4x - 3 = 7$ becomes $4x - 3 + 3 = 7 + 3$

$\Rightarrow 4x = 10$

Similarly, we can subtract the same term from both sides.

Another way of thinking

Operation 4 may be thought of in another way. The rule is:

Move the term across and change its sign

For example:

$3x - 4y = 2x + 7y$

To get xs on one side and ys on the other, move the $4y$ to the right, changing its sign to +, move the $2x$ to the left, changing its sign to:

$3x - 2x = 7y + 4y$

This rule applies only when the terms are joined by + and − signs.

Operation 5 is expressed in another rule:

Move the term across, putting it 'above' if it was 'below' or 'below' if it was 'above'

For example:

$$\frac{5x}{2\pi} = \frac{y}{3}$$

The 2π is below the line; move it across and put it above the line.

$$5x = \frac{2\pi y}{3}$$

This rule applies only when the variables or coefficients being moved are all part of a *single* term. For example in:

$$\frac{5x + 7}{2\pi} = \frac{y}{3}$$

The 7 cannot be moved on its own, as there are *two* terms, $5x$ and 7, above the line on the left. But we can move $(5x + 7)$ together:

$$\frac{1}{2\pi} = \frac{y}{3(5x + 7)}$$

5 Multiply both sides by the same factor.

$3y - 2x = 5$ becomes $2(3y - 2x) = 2 \times 5 = 10$

It is often useful to do this to get rid of a divisor on one side.

$\dfrac{3n + 4}{m} = 5$ becomes $m \times \dfrac{3n + 4}{m} = m \times 5$

$\Rightarrow 3n + 4 = 5m$

If both sides are single fractions, we can *cross-multiply* in one step.

$\dfrac{5x + 4}{y} = \dfrac{4z}{3w}$ becomes $3w \times (5x + 4) = y \times 4z$

$\Rightarrow 3w(5x + 4) = 4yz$

6 Divide both sides by the same factor.

$4f + 20 = \pi$ becomes $\dfrac{4f + 20}{4} = \dfrac{\pi}{4}$

$\Rightarrow f + 5 = \dfrac{\pi}{4}$

7 Square both sides.

$2a = b + 3$ becomes $(2a)^2 = (b + 3)^2$

$\Rightarrow 4a^2 = b^2 + 6b + 9$

Similarly, we can cube both sides or raise to even higher powers.

8 Take the square root of both sides.

$9x^2 = y - 4$ becomes $\sqrt{9x^2} = \sqrt{y - 4}$

$\Rightarrow 3x = \pm\sqrt{y - 4}$

The ± sign indicates that the square root may be positive or negative, mathematically, though there may be practical reasons why it has to be one or the other.

9 Take the reciprocal of both sides.

$7v - 4 = 5$ becomes $\dfrac{1}{7v - 4} = \dfrac{1}{5}$

If one or both sides are single quotients, we can simply turn them both upside down.

10 Take the log both sides.

$4p + 5 = 2q$ becomes $\log(4p + 5) = \log 2q$

Changing the subject

If an equation has a single variable on one side (usually on the left), that variable is the **subject** of the equation. It may happen that we are given an equation in which the variable we want to be the subject is mixed up with several other variables. For example, we may be given the equation for the $-3\,\text{dB}$ point of a filter circuit:

$$f = \frac{1}{2\pi RC}$$

The subject of this equation is the frequency f. We may instead need an equation to calculate the resistance R, given values of f and C. We need to change the subject from f to R. We use the operations listed above, but there is no special order in which they should be applied. The best order of operations depends on the structure of the equation. There are certain things to aim for. We will say what these are as we work this example:

Aim	Operation (see list)	Effect
Move the term containing R to the left	2	$\dfrac{1}{2\pi RC} = f$
Move R above the line	9	$2\pi RC = \dfrac{1}{f}$
Remove $2\pi C$ from the left	6	$R = \dfrac{1}{2\pi fC}$

R is now isolated on the left and becomes the new subject of the equation.

Another example: Given $y = a\sqrt{x} - b$, change the subject to x.

Aim	Operation (see list)	Effect
Move the term containing x to the left	4	$y - a\sqrt{x} = -b$
Remove y (a non-x term) to the right	4	$-a\sqrt{x} = -b - y$
Clear negating signs	3	$a\sqrt{x} = b + y$
Remove a to the right	6	$\sqrt{x} = \dfrac{b + y}{a}$
x is the new subject, but is square-rooted	7	$x = \left(\dfrac{b + y}{a}\right)^2$

The order of the operations could have been different, arriving at the same result. With practice, several of the operations can be done in one step.

An example in which the subject occurs twice on the right:

Given $y = \dfrac{x - 2}{x + 1}$, change the subject to x.

Aim	Operation (see list)	Effect
Move the terms containing x to the left	2	$\dfrac{x - 2}{x + 1} = y$
Remove the divisor to the right	5	$x - 2 = y(x + 1)$
Multiply out the brackets	1	$x - 2 = xy + y$
Move the term containing x to the left	4	$x - 2 - xy = y$
Move the non-x term to the right	4	$x - xy = y + 2$
Factorize to separate x from y	1	$x(1 - y) = y + 2$
Remove the non-x term to the right	6	$x = \dfrac{y + 2}{1 - y}$

Once again, the main aims are to move terms containing the subject to the left, while moving non-subject terms to the right.

An example with the subject as a reciprocal.

Given $\dfrac{1}{x} = 2 + \dfrac{3}{y}$, change the subject to x.

Aim	Operation (see list)	Effect
x is already on the left, but we must sum the terms on the right, ready for operation 9	1	$\dfrac{1}{x} = \dfrac{2y + 3}{y}$
Now take reciprocals	9	$x = \dfrac{y}{2y + 3}$

Success in changing the subject is mainly a matter of practice. Try the questions below.

Test yourself 4.1

Change the subject of each equation to the variable given.

1 $E - IR = V$ to I **2** $V = \dfrac{\pi r^2 h}{3}$ to h

3 $Pg = ma$ to a

4 $mv - 3u = c$ to m

5 $t = \sqrt{2s - k}$ to s

6 $y = ax^2 + b$ to x

7 $2a = \sqrt{\dfrac{x}{y}}$ to y

8 $(y + 3)(y - 3) = 2x + 5$ to y

9 $\dfrac{1}{R} = \dfrac{1}{R_1} + \dfrac{1}{R_2}$ to R

10 $\dfrac{a}{2a + y} = \dfrac{b}{3b + 2y}$ to a

11 $x = \dfrac{5 + 2y}{4 + y}$ to y

12 $S = 2\pi rh + \pi r^2$ to h

Quadratic equations

A quadratic equation is one which has the form:

$$ax^2 + bx + c = 0$$

Because $\sqrt{x}$ can be either positive or negative, this equation has two possible solutions, or **roots**. Occasionally the two roots may be equal, so there is only one solution. A quadratic equation can be written with one or two of its terms on the right side, in which case the zero disappears. For example,

$$ax^2 + bx = c$$

This is converted to the standard quadratic form by moving the c to the left and changing its sign.

Two commonly used ways of solving quadratic equations are:

1 Factorize, then equate the factors to zero.

Example

$$x^2 - 3x - 10 = 0$$

$\Rightarrow \quad (x - 5)(x + 2) = 0 \quad\quad\quad (1)$

$\Rightarrow \quad\quad\quad (x - 5) = 0$

$\quad\quad\quad\text{or} \quad (x + 2) = 0$

$\Rightarrow \quad\quad\quad\quad\quad x = 5$

$\quad\quad\quad\text{or} \quad\quad\quad x = -2$

$\underline{x = -2 \text{ or } x = 5}$

The reasoning behind this technique is that, if the two factors multiplied together equal zero, then the equation is solved when *either one* of the factors equals zero. In the example above, equation (1) is true *either* when $(x - 5)$ equals zero, *or* when $(x + 2)$ equals zero. If $(x - 5)$ equals zero, then x is 5. If $(x + 2)$ equals zero, then x is -2. With either value of x the original equation is satisfied (is balanced). In this way we obtain two solutions, either of which satisfies the equation.

This method is applicable when the left side of the equation factorizes easily, giving roots which are integers.

2 Use the quadratic formula:

Given the equation $ax^2 + bx + c = 0$

then $\quad x = \dfrac{-b \pm \sqrt{(b^2 - 4ac)}}{2a}$

Example
$x^2 - 2x - 7 = 0$

$a = 1 \qquad b = -2 \qquad c = -7$

$\Rightarrow \quad x = \dfrac{-(-2) \pm \sqrt{(-2)^2 - 4 \cdot 1 \cdot -7}}{2 \cdot 1}$

$\quad = \dfrac{2 \pm \sqrt{32}}{2}$

$\Rightarrow \quad x = \dfrac{2 + 5.66}{2} = \dfrac{7.66}{2} = 3.83$ (2 dp)

or $\quad x = \dfrac{2 - 5.66}{2} = \dfrac{-3.66}{2} = -1.83$ (2 dp)

$x = 3.83$ or -1.83 (2 dp)

The expression $-b^2 - 4ac$ is known as the **discriminant**. The value of this tells us something about the roots of the equation:

(i) *Positive discriminant*: The equation has two real (page 4) roots, as in the example above.

(ii) *Zero discriminant*: There are two real *equal* roots.

(iii) *Negative discriminant*: There is the problem of finding the square root of such a discriminant. In the world of real numbers it is not possible to find a value for it. Chapter 14 shows that the square root of the discriminant can be thought of as an imaginary number. A quadratic equation with a negative discriminant has imaginary roots.

Test yourself 4.2

Solve these equations by factorizing, if necessary.

1 $(x + 2)(x - 4) = 0$
2 $(2x - 3)(4x + 5) = 0$
3 $x^2 + 4x - 12 = 0$
4 $x^2 - 4x + 3 = 0$
5 $x^2 - 3x = 10$
6 $x^2 = 9$

Solve these equations correct to two decimal places, by using the quadratic formula.

7 $x^2 - 2x - 7 = 0$ **8** $3x^2 - x - 1 = 0$

9 $4x^2 + 2 = 7x$ **10** $2x^2 = 5 - 4x$

Simultaneous equations

When we have two or more equations, both or all of which are satisfied by the same values of the variables, we call them **simultaneous equations**. A typical pair of simultaneous equations is:

$2x + y = 9$ (1)

$9x - y = 13$ (2)

These are numbered (1) and (2) so that we can refer to them later. Note that, because there are *two* unknown quantities, x and y, we need *two* equations to obtain a solution.

The **elimination method** of solving these equations is to combine them together, so as to eliminate either x or y. Suppose that we decide to eliminate y. To do this, the size of its coefficient (ignoring its sign) has to be the same in both equations. In this example, the coefficient of y is 1 in both equations. Since the coefficients have *opposite* signs, we eliminate y by *adding* the equations together. This is what happens:

Equation (1) is $2x + y = 9$ (1)
Equation (2) is $9x - y = 13$ (2)
Add (1) and (2) $11x = 22$

This has eliminated y, leaving an equation in x which can be solved in the usual way:

$x = 22/11 = 2$

The final step is to find y by substituting the known value of x in equation (1):

$2 \cdot 2 + y = 9$
$\Rightarrow \quad y = 9 - 4 = 5$

We would have obtained the same result by substituting in equation (2); the usual thing to do is to substitute in the equation which is easiest to work out.

Thus the solution is: $x = 2$ and $y = 5$.

In the example above, the coefficients of y in the two equations were of *opposite* sign. y was eliminated by *adding* the equations. In other cases we may find that the two coefficients have the *same* sign (both positive or both negative). If this is so we eliminate the variable by *subtracting* one equation from the other.

Example
Take the equations:

$$2x + y = 3 \qquad (1)$$

$$x - 3y = 19 \qquad (2)$$

Here we have the extra complication that none of the coefficients are the same value. This difficulty is overcome by multiplying *both sides* of one of the equations by a constant factor. Deciding to eliminate x, we leave equation (1) as it is and multiply equation (2) by 2:

$$2x + y = 3 \qquad (1)$$

$$2x - 6y = 38 \qquad (3)$$

The result of multiplying (2) by 2 is labelled (3). Since the coefficient of x has the *same* sign in both equations, we *subtract* one equation from the other to eliminate x.

Subtract (3) from (1)

$$7y = -35$$

$$y = -5$$

Finally, substitute $y = -5$ in equation (1)

$$2x + (-5) = 3$$
$$\Rightarrow \quad 2x = 3 + 5$$
$$\Rightarrow \quad 2x = 8$$
$$\Rightarrow \quad x = 4$$

The solutions are: $x = 4$ and $y = -5$.

In the previous example, instead of eliminating x, we could have decided to eliminate y. This is done by leaving (2) as it is and multiplying (1) by 3. The y terms have opposite signs so y is eliminated by adding the equations. The exact way any pair of equations is tackled is a matter of preference, taking into account the size of the coefficients and their signs. Usually the safest course is to try to keep numbers as small as possible to reduce the likelihood of making mistakes in the arithmetic.

Sometimes the value of the coefficient in one equation is not a simple multiple of the coefficient in the other equation. In such cases we have to multiply *both* equations by different factors so as to make the coefficients equal.

Example

$$2x + 3y = 12 \qquad (1)$$

$$3x - 4y = 1 \qquad (2)$$

To eliminate the terms in x:

Equation (1) × 3 $\Rightarrow \quad 6x + 9y = 36 \qquad (3)$

Equation (2) × 2 $\Rightarrow \quad 6x - 8y = 2 \qquad (4)$

Subtract (4) from (3) $\Rightarrow \quad 17y = 34$

$$\Rightarrow \quad y = 2$$

Substitute $y = 2$ in equation (1)

$$\Rightarrow \quad 2x + 6 = 12$$
$$\Rightarrow \quad 2x = 12 - 6 = 6$$

The solutions are: $x = 3$ and $y = 2$.

Another way of solving simultaneous equations is a variation on the elimination method. We illustrate this **substitution method** by using it to solve an earlier example:

$$2x + y = 3 \qquad (1)$$
$$x - 3y = 19 \qquad (2)$$

Make x the subject of (2):

$$x = 19 + 3y$$

Substitute this value for x in (1):

$$2(19 + 3y) + y = 3$$
$$\Rightarrow \quad 38 + 6y + y = 3$$
$$\Rightarrow \quad 7y = -35$$
$$\Rightarrow \quad y = -5$$

Substitute this value in either equation, say (1):

$$2x + (-5) = 3$$
$$\Rightarrow \quad 2x = 8$$
$$\Rightarrow \quad x = 4$$

The solution is as before: $x = 4$ and $y = -5$.

This method may often be used in cases where the elimination method can not be applied. For example,

$$4x - y = 60 \qquad (1)$$
$$xy = 64 \qquad (2)$$

Since one equation contains terms in x and y but the other has a term in xy, it is not possible to eliminate terms in the usual way. Using the method of substitution, make y the subject of (1):

$$y = 4x - 60$$

Substitute this value in (2):

$$x(4x - 60) = 64$$
$$\Rightarrow \quad 4x^2 - 60x = 64$$
$$\Rightarrow \quad 4x^2 - 60x - 64 = 0$$

Divide by 4: $\quad x^2 - 15x - 16 = 0$

Factorize (if this was not possible we could find the roots by the quadratic formula):

$\Rightarrow \quad (x + 1)(x - 16) = 0$

$\Rightarrow \quad x = -1$

or $\quad x = 16$

Substitute $x = -1$ in (1) $\Rightarrow 4(-1) - y = 60$

$-y = 60 + 4$

$y = -64$

Substitute $x = 16$ in (1) $\Rightarrow 4 \cdot 16 - y = 60$

$-y = 60 - 64$

$y = 4$

There are two solutions: $x = -1, y = -64$ and $x = 16, y = 4$.

The elimination and substitution methods are also applicable when we have three simultaneous equations, with three unknowns. Here is an example, using the substitution method:

$2x + 3y - z = 8$ \quad (1)

$x - y + z = 4$ \quad (2)

$3x + 3y - 2z = 5$ \quad (3)

Make z the subject of (2):

$z = 4 - x + y$

Substitute for z in (1):

$2x + 3y - (4 - x + y) = 8$

$\Rightarrow \quad 2x + 3y - 4 + x - y = 8$

$\Rightarrow \quad 3x + 2y = 12$ \quad (4)

Substitute for z in (3):

$3x + 3y - 2(4 - x + y) = 5$

$\Rightarrow \quad 3x + 3y - 8 + 2x - 2y = 5$

$\Rightarrow \quad 5x + y = 13$ \quad (5)

We now have equations (4) and (5) with terms in x and y only; these are solved as a pair of simultaneous equations. We can use either elimination or substitution to solve these, but will use substitution.

Make y the subject of (5):

$y = 13 - 5x$

Substitute for y in (4)

$$3x + 2(13 - 5x) = 12$$
$$\Rightarrow \quad 3x + 26 - 10x = 12$$
$$\Rightarrow \quad -7x = -14$$
$$\Rightarrow \quad x = 2$$

Substitute for x in (5)

$$10 + y = 13$$
$$y = 3$$

Now we can return to the original equations and substitute for x and y in one of these to find z.

Substituting in (2)

$$2 - 3 + z = 4$$
$$\Rightarrow \quad z = 4 - 2 + 3 = 5$$

The solution is $x = 2$, $y = 3$, $z = 5$.

A third method of solving simultaneous equations is described in Chapter 13.

Test yourself 4.3

Solve these pairs of simultaneous equations.

1. $2x + 3y = 13$
   $-2x + 3y = 5$

2. $-3x + y = -11$
   $6x - y = 23$

3. $4x + 5y = 17$
   $2x - 3y = -19$

4. $x + y = 3$
   $3x - y = 25$

5. $x = y + 1$
   $2y + x = 7$

6. $y = 2x + 11$
   $4x + 3y = -37$

7. $3x - 4y = 7$
   $2x + 5y = -3$

8. $6x - 2y = -11$
   $-4x + 3y = 19$

9. $-4x + 3y - 36 = 0$
   $5x - 2y + 31 = 0$

10. $5.4x - 3.2y = 22.6$
    $6.3x + 1.9y = 15.1$

11. $x - y = -1$
    $4x - y^2 = -1$

12. $xy + 2x = 21$
    $y = x + 2$

Solve these simultaneous equations for three variables.

13. $2x - 3y + z = 13$
    $x + 4y - 2z = -7$
    $5x + y + 3z = 16$

14. $6x + y + z = 0$
    $2x + 2y - 3z = 5$
    $x - 3y - 6z = 3$

Expressing results

Significant figures in a number are those starting with the left-most non-zero digit and finishing with the right-most non-zero digit. The exception is when a number finishes with one or more zeros to the right of the decimal point (see last two examples, and compare with the second example which has no decimal point).

Example	Number of significant figures
3.142	4
5600	2, 3 or 4
0.002 34	3
10.02	4
24.00	4
5.200 00	6

The reason that the zeros are significant in the last two examples is that they imply the *precision* of the number:

$x = 24.00$ implies $23.995 \leq x < 24.005$

$x = 5.200\,000$ implies $5.199\,995 \leq x < 5.200\,005$

The terminal zeros in the second example are not *necessarily* significant; they could be there just to fill the places before the decimal point. If there is doubt, specify the number of significant figures, for example 5600 (2 sf), 5600 (3 sf) or 5600 (4 sf).

Numbers should not be written as if they are more precise than they really are. In practical electronics, results are often significant to no more than three figures. As a general rule:

The maximum number of significant figures in a result can never be more than in the *least* precise value used in the calculations.

Example
A current of 25 mA flows through a precision resistor value 131 Ω, tolerance 0.1%. What is the p.d. across the resistor? The p.d. is calculated as follows. $V = IR = 0.025 \times 131 = 3.275 = 3.3$ V (2 sf). The resistor value is very precise but the current is known only to two significant figures. Therefore there can be no more than two significant figures in the result.

Decimal places

Another way of specifying the precision of a number which has digits after the decimal point is to state the number of decimal places. For example, sin 2 = 0.9093 (4 dp).

Rounding

This is described in the box on page 27.

Scientific form

This is also known as *standard form*. There is one digit before the decimal point and the remaining significant figures after it. The number is multiplied by a power of 10 to give it its correct value.

Example	*Scientific form*
84 600	8.46×10^4
145.6	1.456×10^2
0.000 742	7.42×10^{-4}

Most hand calculators have the ability to work in scientific form.

Engineering form

There are one to three non-zero digits before the decimal place. The number is multiplied by a power of 10 which is always a multiple of 3.

Example	*Engineering form*
84 600	84.6×10^3
145.6	145.6 ($\times 10^0$ understood)
0.000 742	742×10^{-6}

Test yourself 4.4

1. Round to the number of significant figures indicated.
   - **a** 45.6 (2 sf)
   - **b** 3204.73 (5 sf)
   - **c** 35 170 (3 sf)
   - **d** 0.007 64 (2 sf)
   - **e** 6 (2 sf)
   - **f** 798 (1 sf)

2. Round to the number of decimal places indicated.
   - **a** 4.3291 (3 dp)
   - **b** 5.6584 (2 dp)
   - **c** 7.747 (1 dp)
   - **d** 0.007 37 (4 dp)

3. Express in scientific form.
   - **a** 1429
   - **b** 23 470
   - **c** 0.075
   - **d** 0.000 000 1
   - **e** 24.6×10^2
   - **f** 473.6×10^{-7}

4. Express in engineering form the values given in question 3.

As a revision exercise, work the 'Try these first' questions on page 33.

5 Looking at graphs

From the plot of the forward transfer characteristic of a transistor to the trace on an oscilloscope screen, graphs are invaluable in electronics as a means of presenting and analysing data. This chapter outlines the main features of graphs, relating them also to functions.

You need to know about logs (page 25) and equations (Chapter 4, page 33).

Try these first

1 Examine this equation:

$$2y = 3x - 7$$

   a Without plotting the graph, say what you can about the curve that this equation would produce.

   Calculate the value of   b the gradient

                                  c the y-intercept

                                  d the x-intercept

   e Does the point A(4,2) lie on the line of this graph?

   f What is the distance along the line of the graph from point B, at which $x = 2$, to point C, at which $x = 5$?

2 For each of the graphs in Figure 5.1, identify the type of curve and write the equation.

3 If $f(x) = 2x^2 - 3x + 4$, what is the value of

   a $f(5)$

   b $f(0)$

4 Which of the functions A–E listed below is:

   a periodic

   b piece-wise

   c undefined for one value of x (state which value)

   d not a function

   A $f(x) = \dfrac{x + 2}{x - 3}$ **B** $f(x) = 3 + 4 \sin x$

   C $f(x) = 2 + x - \sqrt{3x}$ **D** $f(x) = 4x^2 - 3$

   E $f(x) = 3$ $0 < x < 2$
         $f(x) = 2x - 7$ $2 < x < 3$

LOOKING AT GRAPHS   49

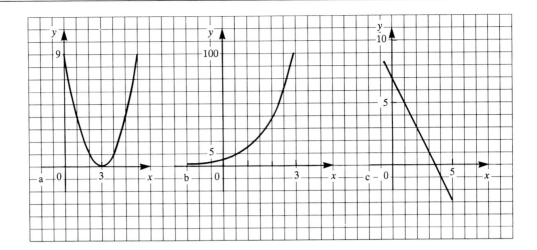

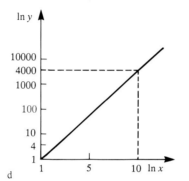

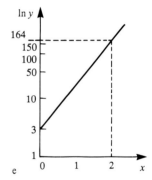

Figure 5.1

Graph variables

An equation is a precise way of describing the relationship between two or more quantities but it is often much clearer to display the equation as a graph. Most of the graphs used in electronics show us the relationship between *two* quantities, or variables; for instance, between time and the charge on a capacitor, or between the base current and the collector current of a transistor. It is possible to draw graphs showing the relationship between three or even more variables, but such graphs are both difficult to draw and hard to interpret. We shall keep to graphs which relate only two variables.

When two variables are shown on a graph, the value of one of them usually depends upon the value of the other. Otherwise, there would be no point in plotting the graph. In the example of the capacitor quoted above, the *charge* on the capacitor depends on the length of *time* for which the capacitor has been charging (or discharging). This is a one-way relationship—*time* cannot depend on how much *charge* there is on the capacitor. We say that time is the **independent variable**, while charge is the **dependent variable**. In the second example, the collector current depends on the base current, not the other way about, so base current is the independent variable and collector current is the dependent variable. Collector current may depend on factors other than the base current, for example on the gain of the transistor, on its temperature or on the collector–emitter voltage, but we assume for the purposes of the graph that all these other factors are being held constant.

Coordinates

We normally think of a graph as a straight or curved line, but really the line is made up of a very large number of individual points. The points of most graphs are so close together that they merge to form a continuous line.

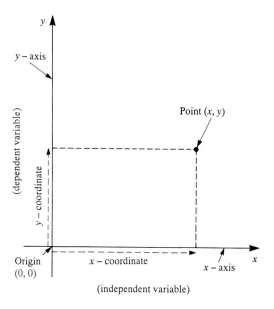

Figure 5.2

Figure 5.2 illustrates the terms used when describing graphs. The location of any point on the graph is specified by its **coordinates**. The coordinates may be stated in two different ways, one of which we shall leave until later (page 93). In this section we deal with **rectangular coordinates**, sometimes known as **Cartesian coordinates**. We specify the coordinates of a point by writing two numbers in brackets, the x-coordinate (the independent variable), followed by the y-coordinate (the dependent variable). For example, if the charge on a capacitor is 2 coulombs at time 3 seconds, the point representing this situation has the coordinates (3,2).

The x-coordinate is sometimes called the **abscissa** and the y-coordinate is sometimes called the **ordinate**.

Although most graphs are continuous lines, it is possible to have a graph made up of only a few distinct and widely-spaced points. For example, we

LOOKING AT GRAPHS 51

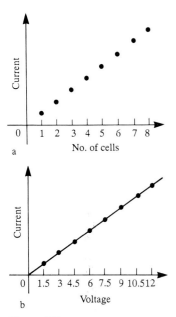

Figure 5.3

could plot the current through a given resistor when it is wired to different numbers of cells connected in series. The number of cells is a whole number, so the graph consists of an individual point for each number of cells tested (Figure 5.3a). This is the same as on the number line for whole numbers (page 4). Strictly speaking, we should not join such points by a continuous line. Fractions of cells, such as 1.45 cells or 4.28 cells, have no meaning, so neither does a line joining the points. However, it is often easier to visualize how voltage varies with the number of cells if a line is drawn to connect the points, so such a line may be allowed, even though it is not mathematically correct.

By contrast, a graph of the current through a given resistor, against the *voltage* produced by different numbers of cells in series can legitimately have its points connected (Figure 5.3b). It is convenient to take the measurements using 0, 1, 2, 3, . . . cells but, if we wanted to, we could obtain intermediate voltages by other means (such as a variable power pack). Voltages such as 1.75 V or 3.83 V *do* have a meaning, even though we may not actually have used these voltages when making the measurements. In this case, the graph represents all the possible voltages over the range tested, and we can read the corresponding currents from the graph.

Straight-line graphs

Many of the graphs we come across in electronics are straight lines. Such a graph is called a **linear** graph. A straight line represents an equation in which both variables are of the first degree. The equation may contain x and y, but not x^2, y^2 or higher powers of these variables.

Given the equation of a linear graph, the line may be drawn in three different ways.

1 *Calculate two or more points* and join them by a straight line.

Example
Graph $y = x + 2$.

If $x = 1$, then $y = 3$ point (1,3)
If $x = 2$, then $y = 4$ point (2,4)
If $x = 3$, then $y = 5$ point (3,5)

Plot the points, join them, and continue the line beyond them in both directions (Figure 5.4). The line is not limited in range to the values you chose for calculating it.

2 *Calculate the y-intercept and gradient*. These terms need to be explained. The **y-intercept** is the point at which the line cuts the *y*-axis. The **gradient** is the ratio:

$$\frac{\text{increase in } y}{\text{increase in } x}$$

Example
Graph $y = 2x + 1$.

The *y*-intercept is where $x = 0$. Substituting this in the equation we find:

$$y = 0 + 1 = 1 \Rightarrow \text{point } (0,1)$$

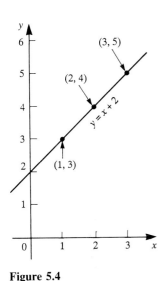

Figure 5.4

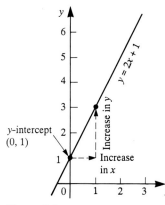

Figure 5.5

The equation shows that, as x increases by 1, y increases by $2x$, that is, by 2.

Mark the point (0,1), then from this point, count along 1 and up 2 to find another point on the line. Join these points and extend the line in both directions (Figure 5.5).

3 *Calculate the x-intercept and y-intercept, and join them.*

Example
Graph $y = 3x - 3$.

The x-intercept is where $y = 0$. Substituting this in the equation, we find:
$$0 = 3x - 3$$
$$\Rightarrow \quad x = 1 \quad \Rightarrow \quad \text{point } (1,0)$$

Calculate the y-intercept as in the previous example:
$$y = 0 - 3 = -3 \quad \Rightarrow \quad \text{point } (0,-3)$$

Join these points and extend the line in both directions (Figure 5.6).

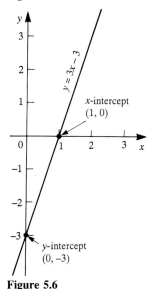

Figure 5.6

Test yourself 5.1

Plot the graphs of these equations, by calculating two or more points.

1 $y = 3x + 2$ **2** $y = x/2 - 4$ **3** $3y = x + 36$

Plot the graphs of these equations, by calculating the y-intercept and the gradient.

4 $y = 4x - 3$ **5** $y = 2 - x$ **6** $4y = 3x + 7$

Plot the graphs of these equations by calculating the x-intercept and the y-intercept.

7 $2y = 5x - 7$ **8** $y = x/2 + 3$ **9** $y = 6 - 2x$

Gradient/intercept equation

The most useful equation for a linear graph (Figure 5.7) is:

$$y = mx + c$$

In this equation, the two key constants are:

c, the y-intercept (when $x = 0$, then $y = c$)
m, the gradient

In the second example above, $c = 1$ and $m = 2$.

If the coefficient of y is 1, then m equals the coefficient of x. If the coefficient of y is other than 1, make the coefficient of y equal to 1 by dividing the equation on both sides by the coefficient of y:

$$4y = 5x + 7$$
$$\Rightarrow \quad y = 1.25x + 1.75$$
$$c = 1.75, \quad m = 1.25$$

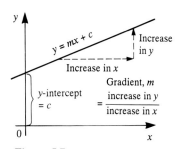

Figure 5.7

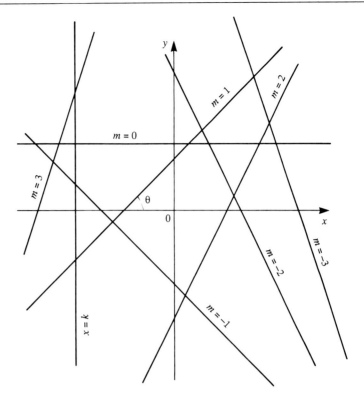

Figure 5.8

Steepness

A positive gradient slopes up to the right. All the examples above have a positive gradient. But, if y decreases as x increases, the 'increase' in y is negative and the gradient is negative. The line slopes down to the right. Figure 5.8 illustrates what various gradients look like, assuming that x- and y-coordinates are plotted on the same scale.

There are two special cases of the gradient:

Line parallel to the x-axis:

$$m = \frac{0}{\text{increase in } x} = 0$$

when $m = 0$ then $y = c$. This equation does not contain x, so the value of y is constant and it is not related to changes in x.

Line parallel to the y-axis:

$$m = \frac{\text{increase in } y}{0}$$

This gives an impossibly 'infinite' gradient. y drops out of the equation. The equation becomes $x = k$ in which k is a constant. The value of x is constant and it is not related to changes in y.

> **A symbol for rate of change**
>
> The gradient of a curve is the rate of change of one variable (y) with respect to the other (x). Instead of representing this by m, we can represent it by:
>
> $$\frac{dy}{dx}$$
>
> This symbol does not mean 'd times y divided by d times x'. If it did, the d's would cancel out! The symbol as a whole stands for the operation of *differentiating*, which we shall describe in detail in Chapter 9, Rates of Change.
>
> In many electronics books this symbol is also used as a shorthand way of specifying a gradient or rate of change, particularly for quantities which change with *time*. Thus dQ/dt represents the rate of change of charge Q (on a capacitor, for example). Another example is dI/dt, representing the rate of change of current. Although quite elementary books often use this symbol as a convenient way of indicating a rate of change, it is rarely that those books actually differentiate any of the quantities involved. As long as you understand what the symbol means, you do not need to know anything about differentiation. Of course, when the theory is taken to greater depths, it then becomes necessary to know what differentiation is. This is when you will need to study Chapter 9.

Provided that the same scale is used on both axes we can express the gradient in another way:

$$m = \frac{\text{increase in } y}{\text{increase in } x} = \tan \theta \text{ (see Figure 5.8)}$$

> **Explore this**
>
> Investigate linear graphs with different c's and m's. Use a graph-plotting program or, at least, use a calculator or micro to calculate points for you to plot on paper. Note what gradients of different values *look* like. Before you plot it, try to predict what the graph of a given equation will look like.

Testing a point

To find if a given point lies on a given line, substitute its coordinates in the equation of the line. If the resulting equation is true, the point is on the line.

Examples
Is the point (1,8) on the line $y = 3x + 5$?

Substituting gives $8 = 3 \cdot 1 + 5 = 8$ True

The point is on the line.

Is the point (7,1) on the line $5y = 2x - 10$?

Substituting gives $5 = 2 \cdot 7 - 10 = -4$ False

The point is not on the line.

Distance between two points

The distance between two points on a linear graph is calculated by using Pythagoras' theorem (Figure 5.9).

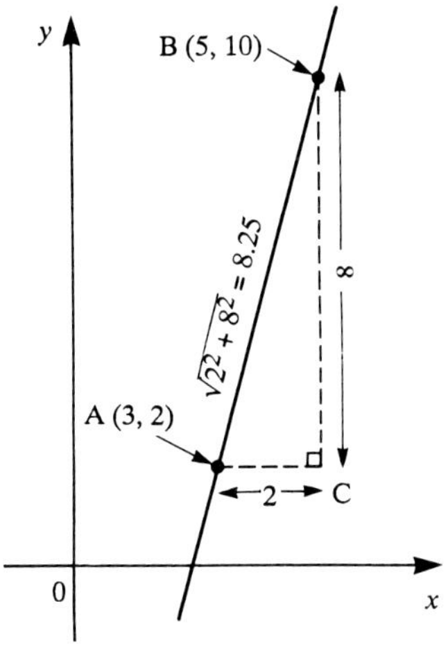

Example
Calculate the distance between point A(3,2) and B(5,10).

The difference of x coordinates is $5 - 3 = 2$

The difference of y coordinates is $10 - 2 = 8$

Sum the squares (distance AB)$^2 = 2^2 + 8^2 = 4 + 64 = 68$

Take the square root   AB = $\sqrt{68}$

AB = 8.25 (2 dp)

Note that we use the *absolute* (positive) differences, and that we take the *positive* square root. Although we have used a drawing to illustrate this technique, there is no need to make a drawing when doing the calculation.

Figure 5.9

Test yourself 5.2

Find (a) the value of c, (b) the value of m, and (c) the equation for each of the graphs in Figure 5.10.

Discover if the point is on the line.

10 $y = 2x + 7$, P(3,13)   **11** $y = 2x + 7$, P(2,9)
12 $3y = 5x - 3$, P(4,5)   **13** $2y + 3x = 5$, P(3,-2)
14 $x + 3y - 3 = 0$, P(2,2)   **15** $y = 2x - 6$, P(1,-3)

Find the distance between the pairs of points, to three decimal places.

16 P(3,2) and Q(4,1)   **17** P(4,-7) and Q(7,-4)
18 L(4,4) and M(-4,-4)   **19** A(5,2) and B(10,-7)

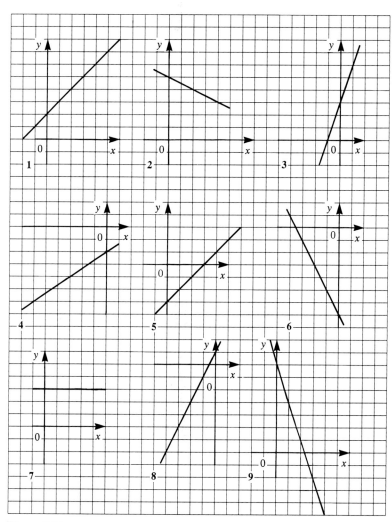

Figure 5.10

Recognizing graphs

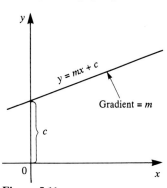

Figure 5.11

It is very useful to be able to recognize a graph to the extent of being able to look at its shape and then say what kind of equation it has. Conversely, it is useful to be able to sketch a graph, given its equation. Here we look at common kinds of graph.

Linear (straight-line) graph: its general equation is:

$$y = mx + c$$

where m is the gradient and c is the y-intercept (Figure 5.11).

Parabola: The simplest equation for a parabola is:

$$y = x^2$$

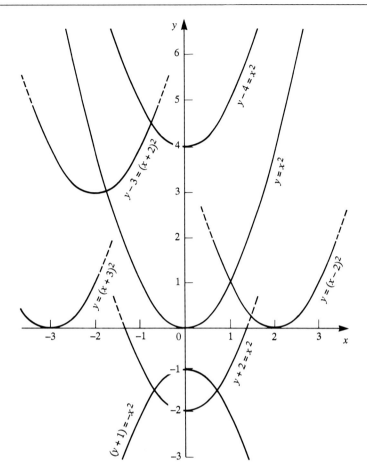

Figure 5.12

This particular parabola is symmetrical about the y-axis (Figure 5.12). The equation may be varied by adding or subtracting constants:

Adding a to the y shifts the curve down a units

$(y + 2 = x^2)$

Subtracting a from the y shifts the curve up a units

$(y - 4 = x^2)$

Adding b to the x shifts the curve left b units

$(y = (x + 3)^2)$

Subtracting b from the x shifts the curve right b units

$(y = (x - 2)^2)$

By adding or subtracting constants to both y and x, the curve is shifted both up or down and left or right. If $y - 3 = (x + 2)^2$, the parabola is shifted 3 up and 2 to the left.

Negating the right side of the equation turns the parabola upside down, as shown by the curve $y + 1 = -x^2$.

Unlike the linear graph, the parabola has a gradient which varies from point to point. It is not possible to measure the gradient of a parabola directly, because there is *no* increase of y or increase of x at a *point* on the graph, so it is impossible to calculate

$$\frac{\text{increase of } y}{\text{increase of } x}$$

Later we shall see ways in which this problem is overcome. For the moment a practical solution is to draw a straight line which *just* touches the parabola at the point for which the gradient is to be found. Such a line is called a **tangent** to that point. We then measure the gradient of the tangent and this has the same value as the gradient of the parabola at that point. Figure 5.13 shows that the gradient of this curve is $m = 2x$.

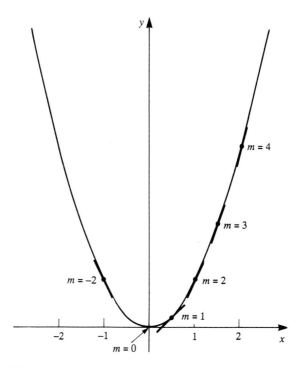

Figure 5.13

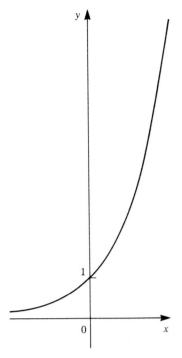

Exponential curve: The curve has the shape shown in Figure 5.14. The simplest equation of a curve of this type is:

$$y = e^x$$

In this equation, e is the exponential constant (page 5). The y-intercept of the curve is 1. The curve has the property that its gradient at any point equals y. That is to say, when $y = 1$ the gradient is 1, when $y = 2$ the gradient equals 2, when $y = 3$ the gradient equals 3, and so on (Figure 5.15). Thus the curve becomes steeper and steeper. This explains why it is sometimes called the **growth curve**. Conversely, as y becomes more and more negative the curve slopes less and less steeply. It never *quite* becomes horizontal and never *quite* meets the x-axis. We say that the x-axis as an **asymptote** of the curve, meaning that it is a line to which the curve gets closer and closer, but never actually touches.

The constants in the exponential equation may be varied as for the parabola, with similar results (Figure 5.16). The dashed line below the curve of $y - 4 = e^x$ illustrates the fact that the curve never quite reaches the value 3. The line $y = 3$ is an asymptote to the curve.

The term 'exponential curve' is used for other kinds of curve with e in their equation, as will be explained on page 63.

Figure 5.14

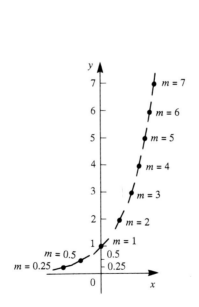

Figure 5.15

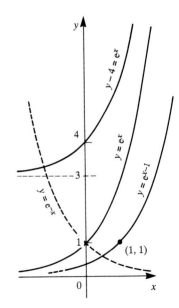

Figure 5.16

Explore this

Use a graph-plotting program, or plot graphs with the help of a calculator or micro, to investigate parabolas and exponential curves. Basic equations are $y + a = \pm(x + b)^2$ for parabolas and $y = ae^{bx}$ for exponential (growth) curves. Try varying the values of the constants a and b (including negative values), and the sign of the right side of the parabola equation. Express the results of your investigations as a set of rules for drawing sketches of curves of these types.

Test yourself 5.3

Identify each graph in Figure 5.17 as linear, parabolic or exponential, and write its equation.

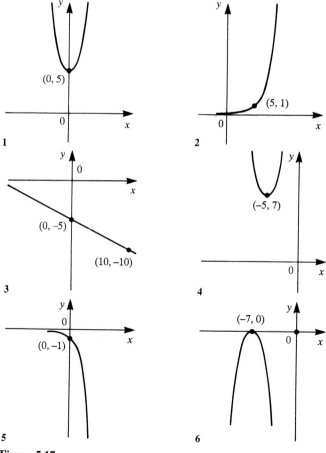

Figure 5.17

Graphs with log scales

There are two main kinds of exponential curve. In one kind, the independent variable x is the index of the exponential constant:

$$y = ae^{bx}$$

This equation is typically used to describe *growth*, either of the money in a bank account, or the growth of an organism such as an oak tree. In electronics, it may describe the increase of electric charge on a capacitor. Compared with the exponential equation given in the previous section, the version above has had two constants a and b included in it. These determine the rate and overall amount of growth.

Here is an example of a series of points based on an equation of this type:

x	2	4	6	8	10	12
y	6.1	24.7	100	406	1645	6671

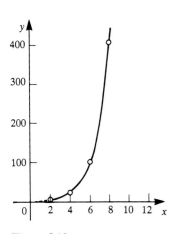

Figure 5.18

The curve (Figure 5.18) slopes up so steeply that there is no room on the page for the last two points. The graph has the general shape of an exponential curve (compare Figure 5.16), but there is no way of being certain that it really is exponential, simply by looking at the graph.

We use logarithms to investigate the curve more closely. The exponential equation above is converted to a different form by taking logs of both sides. We could take logs to base 10, but it simplifies the calculations to take logs to base e:

$$\ln y = \ln a + bx \ln e$$

$\ln e = 1$, giving:

$$\ln y = \ln a + bx$$

This equation is very similar to one we have seen before:

$$y = c + mx$$

This is the equation of the straight-line graph (page 56). The terms on the right are written in a different order, but this does not alter the meaning of the equation. By taking logs of the exponential equation we have obtained an equation for a straight line in which:

— $\ln y$ replaces y, the dependent variable,
— $\ln a$ replaces c, the y-intercept, and
— b replaces m, the gradient.

If we plot a graph with x along the x-axis, as usual, but with $\ln y$ along the y-axis, we should expect to get a straight line with $\ln a$ as its y-intercept and b as its gradient.

Plotting a graph with logs along one axis is a little more complicated than plotting ordinary graphs. One way of doing it is to use a calculator to find the log of the value of y before plotting each point. This is a perfectly good method to use, provided that there are only a few points to plot. If there are many points the task is made easier by using special graph paper. Its lines are evenly spaced along the x-axis, but are spaced logarithmically along the y-axis

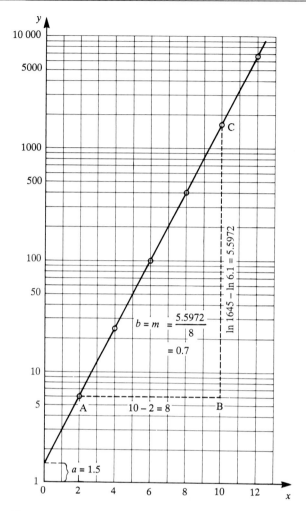

Figure 5.19

(Figure 5.19). This kind of graph paper is called **semi-log** paper or **linear-log** paper.

The great advantage of a semi-log plot is that it is very easy to see whether the line is straight or not. Then we know if the relationship conforms to the exponential equation or not. Checking the exact shape of an exponential curve plotted on ordinary graph paper is virtually impossible.

One point to note about the semi-log plot is that the x-axis is drawn at the value $y = 1$, not at $y = 0$, as on an ordinary graph. This is because logs of zero and negative number do not exist.

Figure 5.18 shows that the semi-log plot of the data above does give a straight line, so we can say that:

The relationship between x and y is exponential.

The next step is to find the values of *a* and *b*. We find *b* by reading off the *y*-intercept, at ln *a*. The *y*-axis is marked in logs, so we read off the value of *a* directly:

$a = 1.5$

To find *b* we need to know the gradient. Having shown that the relationship is exponential, we calculate the gradient. Trying to obtain it from the graph is complicated by the fact that we are using a different kind of scale along the two axes. It is best to calculate it from the original data. We use the values corresponding to points A and C on the graph:

$$b = m = \frac{\text{increase in } y}{\text{increase in } x} = \frac{\ln 1645 - \ln 6.1}{10 - 2} = \frac{5.5972}{8} = 0.7$$

Note that we use logs in the numerator but not in the denominator. The result is:

$b = 0.7$

and so, substituting the values we have found for *a* and *b*:

The equation is $y = 1.5e^{0.7x}$.

Had the graph been based on measured data with an appreciable amount of error, the points might have *approximated* to a straight line, but with none of them close enough to the line to be reliably used in the calculation of *b*. In this case we draw the line which seems to fit the data best and then read the coordinates of two points at the opposite ends of this. These coordinates are used for calculating *b*, as above.

Another relationship which occurs in electronics has *y* depending on a power of *x*. Because *x* has an exponent (page 23) this kind of equation is also called exponential, though it is rather different from the kind we have just discussed. It is not a growth curve, but represents a parabola, a cubic or some similar relationship. A simple equation of this type is:

$y = kx^n$

Given the plot of the results of a test, it is not easy to see if the curve has exactly the right shape to be a parabola, for example, or to discover the exact value of the index. As above, we can plot the curve on special graph paper and obtain a straight line, which we can be sure has the correct shape. Taking logs of both sides of this equation, we find:

$\ln y = \ln k + n \ln x$

We could also use logs to base 10, but we will stay with logs to base e. This log equation has the same form as the straight-line graph, but:

ln *y* replaces *y*, the dependent variable
ln *k* replaces *c*, the *y*-intercept
   n replaces *m*, the gradient, and
ln *x* replaces *x*, the independent variable

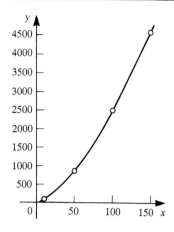

Figure 5.20

In this equation both x and y are replaced by their logs, so the graph is plotted with logs along both axes. We use **log-log** paper.

Here are some data:

x	10	50	100	150
y	79	884	2500	4593

When the graph is plotted on ordinary graph paper (Figure 5.20), it is difficult to recognize any particular curve. When plotted on log-log paper (Figure 5.21) it is clearly a straight line. The value of k is read directly from the graph, at the y-intercept:

$$k = 2.5$$

The gradient may be calculated from two of the data points, as in the previous example. Taking the points where $x = 10$ and $x = 150$:

$$n = m = \frac{\text{increase in } y}{\text{increase in } x} = \frac{\ln 4593 - \ln 79}{\ln 150 - \ln 10} = \frac{4.0628}{2.7081} = 1.5$$

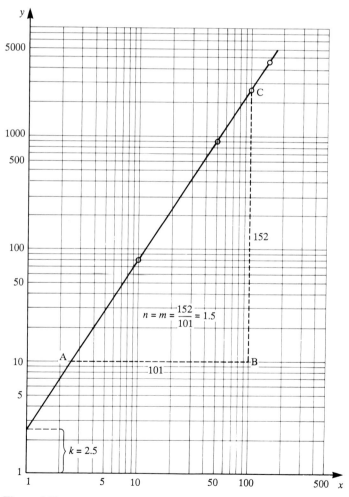

Figure 5.21

Note the logs in the numerator *and denominator.*

$n = 1.5$

substituting the values of k and n:

The equation is $y = 2.5x^{1.5}$.

Since we are using log-log paper, with equal scales on both axes, the gradient may also be found by direct measurement on the graph. Distances AB and BC are measured in millimetres:

$$n = m = \frac{BC}{AB} = \frac{152}{101} = 1.5$$

Test yourself 5.4

1 Confirm by plotting the graph that the data represent an equation of the form $y = ae^{bx}$. Find the values of a and b.

x	1	3	5	7	9
y	3.3	9.0	24.3	66.2	180.0

2 Confirm by plotting the graph that the data represent an equation of the form $y = kx^n$. Find the values of k and n.

x	20	30	40	80	100
y	109	178	251	577	754

3 Find the relationship between x and y, and determine the values of constants involved.

x	2	4	6	8	10
y	5.7	64	265	724	1581

Trig ratios

The three basic trigonometrical ratios express ratios between the sides of a right-angled triangle. They are known as **sine**, **cosine** and **tangent** (see page x). The way that the value of the sine and cosine vary with the size of the angle is best illustrated by drawing graphs.

The graph of:

$y = \sin \theta$

is one that we meet a lot in electronics. It repeats itself every 360°. We say that it is **periodic**, with a period of 360° or 2π radians. Often we leave out the unit when referring to angles in terms of π. It is understood that such an angle is in radians. Thus, $\theta = \pi$ means that $\theta = 3.1416$ radians. It never means that $\theta = 3.1416$ *degrees*.

Another feature of the sine graph is that the value of y has a maximum of 1 and a minimum of −1. We say that its **amplitude** is 1, the amplitude being

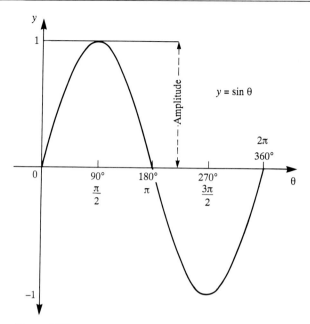

Figure 5.22

the difference between the maximum or minimum value of *y* and its mean value. In electronics, we may also refer to the **peak-to-peak** value of *y*. This is the difference between the maximum and minimum values. The peak-to-peak value is twice the amplitude.

In Figure 5.22 the angle θ is marked both in degrees and radians. Radians are more often used in calculations. When using a calculator it is essential to make sure that it is switched to the correct mode, degrees or radians.

The basic sine equation above may be modified by inserting various constants. Multiplying the whole of the right side by a constant *A* changes the amplitude from 1 to *A* (Figure 5.23). Adding or subtracting a constant *b* on the right raises or lowers the graph, but does not affect the amplitude. Adding a constant angle φ to angle θ shifts the graph to the left. We say that there is a shift of **phase** and that φ is the **phase angle**. Multiplying θ by a constant *n* has the effect of dividing the period by *n*.

Combining these modifications into one equation, we obtain a general equation for the sine curve:

$$y = A \sin(n\theta + \varphi) + b$$

When we are dealing with alternating current, different symbols are used for *n* and θ:

$$y = A \sin(\omega t + \varphi)$$

The constant *b* does not usually appear, unless we are describing an alternating current superimposed on a direct current, value *b*. In this version of the equation, which represents a current fluctuating in time, the independent variable is *t*, the time elapsed since a given instant. The constant

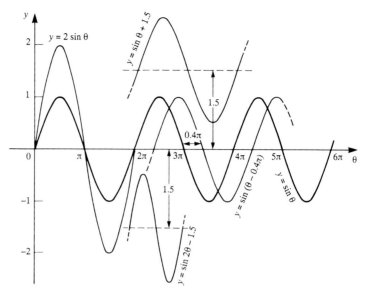

Figure 5.23

ω is the **angular velocity** of the oscillation, which is related to the **frequency** by the equation:

ω = 2πf

Replacing ω by 2πf, we obtain another version of the equation for alternating currents or signals:

y = A sin (2πf + φ)

In this equation, y is the instantaneous current (dependent variable)
A is the amplitude
f is the frequency
φ is the phase angle, if any.

A similar equation is used to express alternating voltages.

The cosine curve, y = cos θ, has the same shape as the sine curve. Like the sine curve, its amplitude is 1 and its period is 360°. It differs from the sine curve in its phase, being 90° ahead of the sine curve, as shown in Figure 5.24.

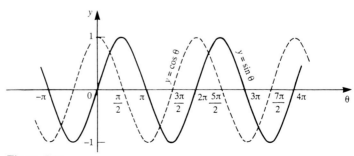

Figure 5.24

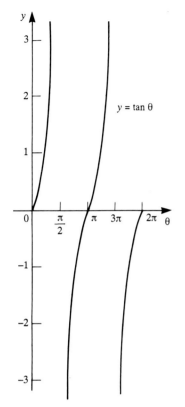

Figure 5.25

The tangent curve is very different in appearance (Figure 5.25) changing sign abruptly at 90° (= π/2) and 270° (= 3π/2). In other words the lines θ = 90° and θ = 270° are asymptotes (page 59) to this curve. The curve is periodic, with period 180°.

As we go through the various **quadrants** (or quarters) of a circle, the signs of the trig ratio are either positive or negative according to a regular pattern. The box summarizes this.

Trig signs					
Quadrant	Angles		sin	cos	tan
	degrees	radians			
1st	0 to 90	0 to π/2	+	+	+
2nd	90 to 180	π/2 to π	+	–	–
3rd	180 to 270	π to 3π/2	–	–	+
4th	270 to 360	3π/2 to 2π	–	+	–

Because the sides of a right-angled triangle have related lengths, depending on the angle θ, the trig ratios are related to each other in a number of ways. These identities are true for all values of θ. The box lists a few of the more important trig identities.

Trig identities

$\sin \theta \equiv \cos(\theta - 90°)$

$\cos \theta \equiv \sin(\theta - 90°)$

$\dfrac{\sin \theta}{\cos \theta} = \tan \theta$

$\sin^2 \theta + \cos^2 \theta \equiv 1$

$\sin(A + B) \equiv \sin A \cos B + \cos A \sin B$

$\sin(A - B) \equiv \sin A \cos B - \cos A \sin B$

$\cos(A + B) \equiv \cos A \cos B - \sin A \sin B$

$\cos(A - B) \equiv \cos A \cos B + \sin A \sin B$

$2 \sin A \cos B \equiv \sin(A + B) + \sin(A - B)$

$2 \cos A \sin B \equiv \sin(A + B) - \sin(A - B)$

$2 \cos A \cos B \equiv \cos(A + B) + \cos(A - B)$

$2 \sin A \sin B \equiv \cos(A - B) + \sin(A + B)$

Explore this

We are not giving *proofs* for the trig identities here, but use a calculator or micro to confirm that the identities hold true for a wide range of values of A and B.

Other graphs

The most important graphs in electronics are the straight line, the parabola, the exponential graph and the sine and cosine graphs. Occasionally we may come across other kinds of graph. Figure 5.26 helps you identify these from their general shape. The hyperbola shown in the figure has the x-axis and y-axis as its asymptotes and, since these are at right-angles to each other, is more specifically known as a **rectangular hyperbola**.

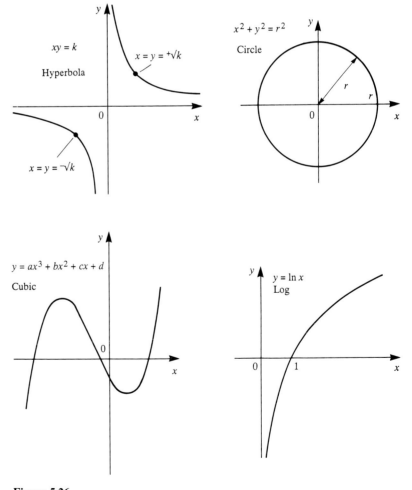

Figure 5.26

Important points

Certain points on a curve are of particular interest. These include:

Stationary point: a point at which the gradient of the curve is zero (Figure 5.27). There are two special kinds of stationary point, a **maximum**, where gradient m is zero as it changes from positive to negative, and a **minimum** where m is zero as it changes from negative to positive. Thus there is a *change of sign* of m at a maximum or minimum.

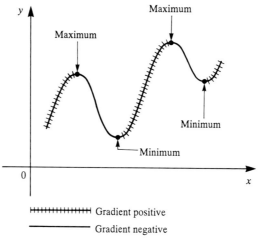

Figure 5.27

Point of inflection: a point where there is a change of curvature. The curve changes in shape from being concave (as seen from below), to being convex, as illustrated by point A in Figure 5.28. Or the curve may change from convex to concave as at point B. There is no change of sign of m at a point of inflection. It may happen that the gradient is zero at a point of inflection, as at C. If so, the point of inflection is also a stationary point.

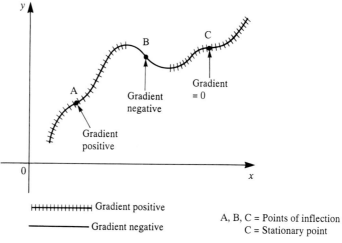

Figure 5.28

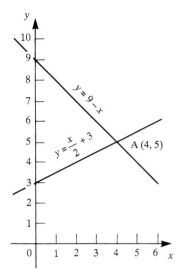

Figure 5.29

Point of intersection: a point where two curves cross each other. Each curve represents a different equation but, at the point of intersection, the values of *x* and *y* are such that they satisfy both equations *at the same time*. In Figure 5.29 we have the line for the equations:

$$y = x/2 + 3 \quad \text{and} \quad y = 9 - x$$

The lines cross at point A(4,5). This is the same thing as saying that the equations are simultaneous equations (page 41) and that their solution is $x = 4, y = 5$.

Plotting the graphs of two simultaneous equations and finding the point (or points) of intersection is one way of solving the equations. We do not usually solve ordinary linear equations in this way, as graphs are subject to errors in drawing and in reading the coordinates of points. Exact results are obtained by using algebra, as on pages 41–45. But complicated equations may not be solvable by easy algebra and plotting the graphs may be the better technique. For example, take these equations:

$$y = x^2 - 2x + 2 \qquad (1)$$

$$y = x^3 - 5x^2 - 4x + 22 \qquad (2)$$

Their curves are a parabola and a cubic curve (Figure 5.30), with three points of intersection, corresponding to the three solutions. To check point B, where

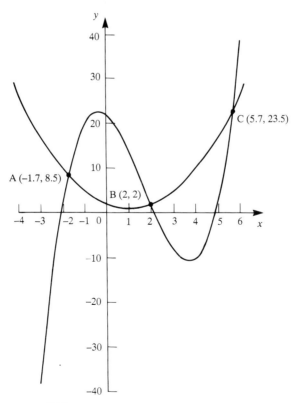

Figure 5.30

$x = 2$ and $y = 2$, substitute these values in equation (1):

$2 = 2^2 - 2(2) + 2 = 4 - 4 + 2 = 2$ True

Also substitute in equation (2):

$2 = 2^3 - 5(2)^2 - 4(2) + 22 = 8 - 20 - 8 + 22 = 2$ True

Both equations are true for point (2,2), so this point lies on *both* lines and satisfies both equations at the same time. The same applies to points A and C, which are two further solutions to these equations.

Functions

A **function** is an expression, the value of which depends on the value of a variable or variables. We often plot a graph to make the action of a given function easier to understand.

For example, the area A of a circle radius r is given by the function:

$$A = \pi r^2$$

The function is written out as an equation. Given any value of the independent variable r, we can calculate A by first squaring r, then multiplying it by π. The equation tells us *what to do* to obtain A, given a value of r. We say that the area is a *function of* the radius. For each different value of r we obtain one and only one value of A. This one-to-one relationship is an essential feature of a function.

Another example of a function is the instantaneous value of an alternating current, as given by the function:

$$i = A \sin 2\pi ft$$

If frequency f and amplitude A are constant, we obtain one and only one value of i for every value of t. Here, i is a function of t.

In general, a function is a mathematical operation or series of operations performed on a given set of numbers to produce another set of numbers. It may be connected to some physical process, as in the second example, but may be entirely mathematical, as in the first one. The set of numbers the function operates on is called the **domain** of the function. The set of numbers resulting from the function is its **range**.

Functions can be written out as equations, as above, or in other ways. One way commonly used it to replace the 'y' by the symbol '$f(\ )$'. In the bracket we write the symbol of the variable, representing the members of the domain. For example, we could re-write the functions above as:

$$f(r) = \pi r^2$$

$$f(t) = A \sin 2\pi ft$$

This way of writing out the function does *not* mean 'f times the symbol in brackets'. The symbol in brackets is known as the **argument** of the function. With most functions we can replace the symbol by any real number that is in the domain, to indicate a particular instance of the function. Taking the first function as an example, we can say:

$$f(3) = \pi \cdot 3^2 = 28.27 \quad (2 \text{ dp})$$

For the same function:

$$f(5) = \pi \cdot 5^2 = 78.54 \quad (2 \text{ dp})$$

The argument does not have to be an integer:

$$f(2.57) = \pi(2.57)^2 = 20.75 \quad (2 \text{ dp})$$

Although all functions may be defined by writing out one or more equations, not all equations define functions. For an equation to represent a function there must be one and one only value of the function for any given value of the argument. For example, the equation of Figure 5.25 defining the circle gives *two* values of the function for every value of *x*:

$$f(x) = +\sqrt{r^2 - x^2} \quad \text{and} \quad f(x) = -\sqrt{r^2 - x^2}$$

For this reason it is not a function. A simple test for a function is to draw a line through it parallel to the *y*- axis. If the line cuts the graph at two or more points, it is not a function. A line through the circle of Figure 5.26 cuts the circle twice. A parabola with equations such as those in Figure 5.12 is a function. However, the curve for $x = y^2$ also gives a parabola (Figure 5.31). This is not a function because a line drawn parallel to the *y*- axis cuts it twice.

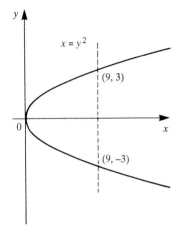

Figure 5.31

Periodic functions

Functions such $f(x) = \sin x$, $f(x) = \cos x$ and $f(x) = \tan x$ repeat the values of *y* after an interval, known as the **period** of the function. For $f(x) = \sin x$ and $f(x) = \cos x$, the period is 2π (Figure 5.24). For $f(x) = \tan x$, the period is π (Figure 5.25). Note that a function does not necessarily have a value for every value in its domain. $y = \tan x$, for example, is indeterminate when $x = 0$.

Piecewise functions

Different equations may be used to define a function over different parts of its domain. An example of a piecewise function is shown in Figure 5.32. We define this function by listing the equation for the different parts of its domain.

$$f(x) = x + 5 \quad x < 0$$
$$f(x) = 5 \quad 0 < x < 5$$
$$f(x) = -2 \quad 5 < x$$

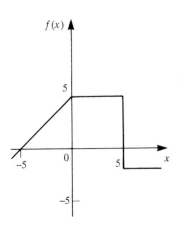

Figure 5.32

Explore this

Investigate the shapes of different functions. Use a graph-plotting program or plot graphs using a calculator or micro to calculate the values. Here are some examples to get you started:

$f(x) = a \sin x + b \sin 2x \qquad -4\pi < x < 4\pi$

Use various values of a and b. Try adding a third term, $c \sin 3x$

$f(x) = \tan^{-1} x \qquad -10\pi < x < 10\pi$

Some computers refer to $\tan^{-1}$ as *arctan* and use the keyword ATN.

$f(x) = (x + a)(x + b)(x + c) \qquad -10 < x\ 10$

Vary the constants a, b and c between -10 and 10. What do you notice about the x-intercepts?

Test yourself 5.5

1 Use a calculator to find the values of the following, to 4 dp.
   a $\sin \pi$
   b $\cos 75°$
   c $\sin 3\pi/2$
   d $\tan 135°$
   e $\tan 306°$
   f $\cos 2.5$ rad
   g $\sin^{-1} 0.75$
   h $\cos^{-1} -0.66$
   i $\tan^{-1} 2$

2 Solve these pairs of simultaneous equations by plotting their graphs.
   a $y = x + 3$
   $y = 3x - 3$
   b $y = x^2 + 2$
   $y = 5x - 2$

3 Calculate the values of the functions for the arguments given, to 4 sf.
   a $f(x) = 2x^2 - 3$, for $f(0)$, $f(2)$ and $f(10)$
   b $f(x) = 3e^{2x}$, for $f(-2)$, $f(0)$, and $f(6)$
   c $f(x) = 3 \sin (x/2)$ for $f(1)$, $f(2.5)$, $f(12)$, all angles in radians

4 Which of the following are functions of x?
   a $y = x$
   b $x^2 + y^2 = 40$
   c $2y = 3x + 6$
   d $xy = 6$
   e $y = |2x|$
   f $y = \cos x$
   g $y = x^2 + 5$
   h $x = \cos y$

6 From theory to practice

> There is no new maths in this chapter. Instead, we take the x's and the y's of the earlier chapters and replace them with V's, I's and other electronics symbols. This will help you understand how maths is used to solve some of the simpler types of electronics problem. The references at the end of each solution tell you where to look for fuller explanations.

P.d. across a resistor

Problem

Given the Ohm's Law formula $I = V/R$, calculate the voltage across a 33 Ω resistor when a current of 0.25 A is flowing through it (2 sf).

Solution

The formula has I as its subject, so we must change the subject to V. Exchanging sides gives:

$$V/R = I$$

Multiplying *both* sides of the equation by R gives:

$$V = IR$$

Now we substitute the values given for I and R:

$$V = 0.25 \times 33 = 8.3$$

The voltage across the resistor is 8.3 V (2 sf).

Comments

(1) The result of the calculation is 8.25 but, since I and R are given to only 2 sf, there is no point in expressing the result in more than 2 sf. It is rounded to 8.3.

(2) The calculation is done using *numbers*. The units, if any, are to be included after the calculation is complete. The unit in which the result is expressed depends on the units in which the variables I and R are expressed. In this case, *amps* × *ohms* gives a result in *volts*.

Reference pages: 34, 46.

Resistance

Problem

Given the Ohm's Law formula $I = V/R$, calculate the resistor required to produce a current of 25 mA when the voltage across the resistor is 12 V (2 sf).

Solution

The formula has I as its subject, so we must change the subject to R. Multiplying both sides by R brings R to the left:

$$IR = V$$

Dividing both sides by I gives:

$$R = \frac{V}{I}$$

Now we substitute the values given for V and I:

$$R = \frac{12}{0.025} = 480 \text{ (2 sf)}$$

The required resistance is 480 Ω (2 sf).

Comments

See comment (2) of the previous solution. The unit of volts/amperes is the ohm. In most circuits, a 470 Ω resistor, which is a standard resistor of the E12 range, would be sufficiently close to the required value.

Reference pages: 34, 46.

Current through a resistor

Problem

Given the Ohm's Law formula $I = V/R$, calculate the current through a 820 Ω resistor when a p.d. of 6 V is applied across it (2 sf).

Solution

Substituting the values given for V and R:

$$I = \frac{6}{820} = 0.0073$$

The current through the resistor is 0.0073 A (2 sf).

Comments

The result 0.0073 A is better expressed in engineering form as 7.3×10^{-3} A. Alternatively, express it in a smaller unit, 7.3 mA.

Resistors in parallel

Problem

The formula for calculating R, the resistance of two or more resistors in parallel is:

$$\frac{1}{R} = \frac{1}{R_1} + \frac{1}{R_2} + \frac{1}{R_3} + \ldots + \frac{1}{R_n}$$

where $R_1, R_2, R_3 \ldots R_n$ are the individual resistances. Calculate the resistance of two resistors, 56 Ω and 120 Ω, wired in parallel (2 sf).

Solution

With many resistances in parallel we take the reciprocals of the resistances, sum them, and take the reciprocal of the sum. With only two resistances, which is probably the situation most often met with, the formula can be simplified.

$$\frac{1}{R} = \frac{1}{R_1} + \frac{1}{R_2}$$

To add the fractions on the right, first find the LCM of R_1 and R_2, which is $R_1 R_2$. Using this as the denominator, convert the right side to equivalent fractions:

$$\frac{1}{R} = \frac{R_2}{R_1 R_2} + \frac{R_1}{R_1 R_2}$$

$$= \frac{R_2 + R_1}{R_1 R_2}$$

Inverting both sides of the equation:

$$R = \frac{R_1 R_2}{R_1 + R_2}$$

Substituting the values given:

$$R = \frac{56 \times 120}{56 + 120} = \frac{6720}{176} = 38$$

The parallel resistance is 38 Ω.

Comments

(1) This is a handy formula.
(2) As a check on the working, the parallel resistance is always less than either of the two resistances. If one resistance is much bigger than the other, the parallel resistance is close in value to the smaller resistance.

Currents in a resistor network

Problem

Find the current I flowing through resistor R_2 in Figure 6.1 (to 2 sf).

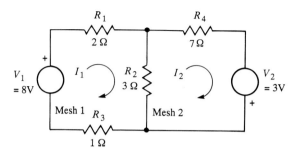

Figure 6.1

Solution

We use the method of mesh current analysis, applying Kirchhoff's Voltage Law (KVL). The fictitious currents flowing in the meshes of the circuit are I_1 and I_2, flowing clockwise as indicated in the figure. Since I_1 and I_2 flow through R_2 in opposite directions, the current through R_2 is the difference:

$$I = I_1 - I_2.$$

We begin by calculating I_1 and I_2. In mesh 1, according to KVL:

Total voltage drops across resistors = emf of the cell

$$(R_1 + R_2 + R_3)I_1 - R_2 I_2 = V_1$$

The p.d. across R_2 due to I_2 is subtracted, as I_2 flows *anti*clockwise through this part of mesh 1.

Similarly, for mesh 2:

$$-R_2 I_1 + (R_2 + R_4)I_2 = V_2$$

Here I_1 is flowing *anti*clockwise through the R_2 part of mesh 2.
Substituting the known values of resistors and voltage sources:

$$6I_1 - 3I_2 = 8 \qquad (1)$$

$$-3I_1 + 10I_2 = 3 \qquad (2)$$

These are *simultaneous equations*, with variables I_1, and I_2, and may be solved by the elimination method.

Equation (1) is $\qquad\qquad 6I_1 - 3I_2 = 8$

Equation (2) × 2 $\Rightarrow$ $\underline{-6I_1 + 20I_2 = 6}$ $\qquad (3)$

Add (1) and (3) $\qquad\qquad 17I_2 = 14$

$\Rightarrow \qquad I_2 = 14/17$

Keep the improper fraction for the moment. Substitute $I_2 = 14/17$ in equation (1):

$$6I_1 - 3 \cdot 14/17 = 8$$
$$\Rightarrow \quad 6I_1 - 42/17 = 8$$
$$\Rightarrow \quad 6I_1 = 8 + 42/17$$
$$= \frac{136 + 42}{17} = 178/17$$
$$\Rightarrow \quad I_1 = 178/102 = 89/51$$

Having solved the equations, the solutions may be converted to decimal form, though with more significant figures than we shall eventually need.

$$I_1 = 89/51 = 1.7451$$
$$I_2 = 14/17 = 0.8235$$

We are now able to calculate the current I through R_2, remembering that I_1 and I_2 are opposite in direction as they pass through this resistor:

$$I = I_1 - I_2 = 1.7451 - 0.8235 = 0.9216$$

The current is 0.92 A in the direction of I_1 (2 sf).

Comment

It is better to keep the improper fractions until the equations are solved, because converting the first one introduces a slight error which will affect the value obtained for the second. When we convert them later, we retain more significant figures than are needed at the end. If we round to 2 sf at this stage we get $I_1 = 1.7$ and $I_2 = 0.8$. This gives $I = 1.7 - 0.8 = 0.9$. This has the same value as the correct answer but has only 1 sf. I could then have any value between 0.85 and 0.95. By working with more significant figures until the very end of the calculation we get a more precise result.

Reference pages: 41, 46.

Potential divider

Problem

The circuit of Figure 6.2 is used to step down the supply voltage to 10 V. R_A and R_B represent the two parts of the track of a variable resistor, with the wiper at point P. If the resistor is adjusted to produce 10 V at P, what is the resistance r of R_B (2 dp)?

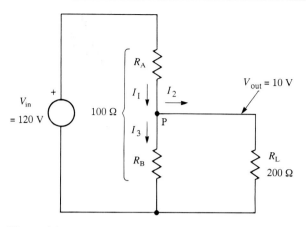

Figure 6.2

Solution

The simplest solution ignores the current being drawn from the potential divider and passing through the load resistor R_L. If the total resistance of R_A and R_B is R, the formula is:

$$V_{OUT} = V_{IN} \times \frac{r}{R}$$

Making r the subject of the equation:

$$r = \frac{V_{OUT} \times R}{V_{IN}}$$

Substituting values given:

$$r = \frac{10 \times 100}{120} = 8.33$$

The resistance of R_B is 8.33 Ω

If R_L is high compared with the resistance of the variable resistor, this calculation gives a result close enough for many purposes. However, if R_L is low, its resistance in parallel with R_B causes a significant drop in the voltage at P. The problem is solved by applying Kirchhoff's Current Law (KCL), which states that the total current entering a node of a network equals the total current leaving it.

$$\text{Current entering P is } I_1 = \frac{110}{100 - r}$$

$$\text{Currents leaving P are } I_2 = \frac{10}{200} = 0.05$$

$$\text{and } I_3 = \frac{10}{r}$$

Applying KCL:
$$I_1 = I_2 + I_3$$
$$\Rightarrow \quad \frac{110}{100-r} = 0.05 + \frac{10}{r}$$

Add expressions on the right, with r as the denominator:
$$\Rightarrow \quad \frac{110}{100-r} = \frac{0.05r + 10}{r}$$

Cross-multiply:
$$\Rightarrow \quad 110r = (100-r)(0.05r + 10)$$
$$= 1000 - 5r - 0.05r^2$$

Collecting all terms on the left, inverting signs as necessary:
$$0.05r^2 + 115r - 1000 = 0$$

This is a quadratic equation with coefficients:
$$a = 0.05 \quad b = 115 \quad c = -1000$$

Solving this by using the quadratic formula:
$$r = \frac{-115 + \sqrt{115^2 - 4 \times 0.05 \times -1000}}{2 \times 0.05}$$
$$= \frac{-115 + \sqrt{13\,425}}{0.01} = \frac{-115 \pm 115.866}{0.1}$$

Only the positive root is applicable here, so:
$$r = \frac{0.866}{0.1} = 8.66$$

The resistance of R_B is 8.66 Ω.

Comment

Increasing R_B compensates for having R_L in parallel with it; their parallel resistance is 8.3 Ω, the value calculated when ignoring I_2.

Reference pages: 16, 34, 40.

Flux density

Problem

A magnetic field with flux Φ and cross-sectional area A has a flux density $B = \Phi/A$. If there is a magnetic field of flux 3.5 mWb with an area of 14 cm², calculate its flux density.

Solution

In the formula, Φ is in weber, A is in square metre and so B is in weber per square metre, usually referred to as tesla. The data are given in other units, milliweber and square centimetre, so we must convert these to the correct units before applying the formula.

$$\Phi = 3.5 \text{ mWb} = 3.5 \times 10^{-3} \text{ W}$$

$$A = 14 \text{ cm}^2 = 14 \times 10^{-4} \text{ m}^2$$

Now substitute these in the formula, but write the powers of 10 as a separate fraction:

$$B = \frac{3.5}{14} \times \frac{10^{-3}}{10^{-4}}$$

Evaluate the fractions separately; the powers of 10 are simplified by using the rule for division of indexes:

$$B = 0.25 \times 10^{-3-(-4)}$$

$$= 0.25 \times 10^1 = 0.25 \times 10 = 2.5$$

The flux density is 2.5T.

Decibels

Problem

(i) An amplifier has a power gain of 80 times. Express this in decibels.
(ii) The amplifier is connected in cascade to a second amplifier with a power gain of 20dB. What is the total power gain of the system?
(iii) If the output of the system is 0.5 mW, what is the input to the system in mW (2 dp)?
(iv) Express this in dBm.

Solution

(i) If input power is P_1 and output power is P_2:

$$\text{gain} = 10 \log_{10} \frac{P_2}{P_1}$$

In this problem $P_2/P_1 = 80$, and:

$$\text{gain} = 10 \log 80 = 10 \times 1.903 = 19 \text{ dB}$$

(ii) The power gain of the two amplifiers together is the sum of their gains in decibel $= 19 + 20 = 29$ dB.

(iii) The power ratio P corresponding to 29dB is calculated from

$$29 = 10 \log P$$

$$\Rightarrow \quad 2.9 = \log P$$

Taking antilogs of both sides

$$794 = P$$

The power gain is 794 times. If the output is 0.5 mW, the input is 0.5/794 = 0.000 63 mW (2 sf).

(iv) The input is 1/10 of 1 mW so:

$$\text{gain} = 10 \log 0.00063$$

To avoid negative signs with gains less than 1 (power *loss*), take the log of the reciprocal and change the sign:

$$\Rightarrow \quad \text{gain} = -10 \log 1\,588$$

$$= -32 \text{ dB}$$

The input power is 32 dB *below* 1 mW. In other words, the input is −32 dBm.

Comment

With decibel calculations we always use logs to base 10, not natural logs. Calculations involving voltage or current gain or loss are similar except that the coefficient 20 is used instead of 10. This is because power is proportional to the *square* of the voltage or current, so the logs have to be *doubled*.

Induced emf

Problem

What emf is induced in a conductor (O, Figure 6.3) which moves across a magnetic field, flux 0.75 µWb, 40 mm square, at an angle of 80° to the field, and at a velocity of 25 m s^{-1} (3 sf)?

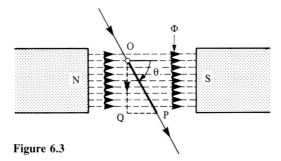

Figure 6.3

Solution

The induced emf is given by:

$$e = B\ell v$$

where B is the flux density, ℓ is the length of wire in the field, and v is its velocity perpendicular to the field.

First calculate B, given by flux divided by cross-sectional area of the field.

$$B = \frac{0.75 \times 10^{-6}}{(40 \times 10^{-3})^2}$$

Separating out the powers of 10:

$$B = \frac{0.75}{40^2} \times \frac{10^{-6}}{(10^{-3})^2}$$

Using the rules for indices:

$$B = \frac{0.75}{1600} \times \frac{10^{-6}}{10^{-6}}$$

The powers of 10 cancel out completely, and we now have to evaluate:

$$B = \frac{0.75}{1600} = 0.000\ 468\ 75 = 4.6875 \times 10^{-4}$$

In Figure 6.3 the vector **OP** represents the motion of the conductor, which is $25\ \text{m s}^{-1}$. The component of this vector perpendicular to the field is:

$$v = 25 \sin 80° = 25 \times 0.9848 = 24.620$$

The induced emf is:

$$e = 4.6875 \times 10^{-4} \times 40 \times 10^{-3} \times 24.620$$
$$= 4616.25 \times 10^{-7}$$
$$= 461 \times 10^{-6}$$

At the final stage, we rounded to three significant figures and also converted to engineering form. The emf is 461×10^{-6} V, better expressed as 461 μV.

Reference pages: 24, 28, 47.

Alternating current

Problem

An alternating current is represented by the equation:

$$i = 12 \sin 785t$$

What are the peak current, the rms current, the frequency, and the period (3 sf)?

Solution

Compare the given equation: $\quad i = 12 \sin 785t$
with the standard equation: $\quad i = A \sin (2\pi f t + \varphi)$

A corresponds to the coefficient 12.

    Peak current = 12.0 A
    rms current = 0.707 × 12 = 8.49 A

$2\pi f$ corresponds to the value 785

    $f = 785/2\pi = 125$ Hz
    Period = $1/125 = 8.00 \times 10^{-3}$ s
          = 8.00 ms

There is nothing to correspond to φ, so there is no phase lag or lead.

Reference pages: 5, 66.

Phasors

Problem

Using a phasor diagram, find the sum of the two alternating voltages represented by $v_1 = 60 \sin 500t$ and $v_2 = 40 \sin (500t + \pi/5)$.

Solution

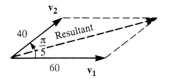

Figure 6.4

Both voltages have the same frequency, so they can be summed by vector addition of their phasors. The phasor for v_1 has no phase lead or lag, so is drawn horizontally, pointing right (Figure 6.4). Its length is scaled to represent a peak amplitude of 60 V. The phasor for v_2 leads v_1 by $\pi/5$, so this is drawn as shown, rotated $\pi/5$ (= 36°) anticlockwise of v_1, and with a length representing 40 V.

Completing the parallelogram, the length of the resultant is equivalent to 95 V. The phase angle of the resultant is 13.75° (= 0.24 rad).

The equation for the sum of these voltages is:

    $v = 95 \sin (500t + 0.24)$

Comment

    1 rad = $(180/\pi)° = 57.30°$ (4 sf).

Reference page: 30.

Inductive impedance

Problem

An alternating voltage with the equation $v = 100 \sin 1600\pi t$ is applied to a coil with inductance 0.2 H, and resistance 500 Ω. Calculate the impedance of the circuit and the current it draws.

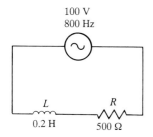

Solution

By comparison with the standard equation, it is seen that the peak voltage is 100 V and the frequency is $1600\pi/2\pi = 800$ Hz. Figure 6.5 shows the circuit and the phasor diagram.

At 800 Hz the reactance of the coil is:

$$X_L = 2\pi f L = 2\pi \times 800 \times 0.2 = 1005 \; \Omega$$

The figure is drawn to scale, with phasors V_R and V_L scaled according to their impedances. The resultant phasor is proportional to the impedance of the whole circuit, the vector sum of the resistance and reactance of the coil. Calculate this using Pythagoras' theorem.

$$Z = \sqrt{(500^2 + 1005^2)} = 1123 \; \Omega$$

The current is $V/Z = 100/1123 = 0.089$ A

The phase angle, $\varphi = \tan^{-1} \dfrac{X_L}{R} = \tan^{-1} \dfrac{1123}{500}$

$$= \tan^{-1} 2.246 = 1.15 \text{ rad} \quad (= 66°)$$

As shown in the diagram, the current lags V by this angle.

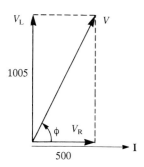

Figure 6.5

Reference pages: 24, 29, 66.

Capacitative impedance

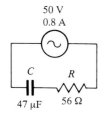

Problem

A capacitor 47 μF and resistor 56 Ω are wired in series across an alternating supply of rms voltage 50 V, and draw an rms current of 0.8 A. What is the frequency of the supply, and what are the voltages across the capacitor and resistor? State the equations for the total voltage and current.

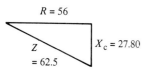

Solution

The impedance of the circuit is:

$$Z = \frac{\text{supply voltage}}{\text{current}} = \frac{50}{0.8} = 62.5 \; \Omega$$

The capacitative impedance is obtained by solving the triangle (Figure 6.6):

$$X_c = \sqrt{62.2^2 - 56^2} = \sqrt{770.25} = 27.75 \; \Omega$$

We can now calculate the frequency:

$$f = \frac{1}{2\pi C X_c} = \frac{1}{2\pi \times 47 \times 10^{-6} \times 27.75} = 122 \text{ Hz}$$

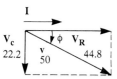

Figure 6.6

Given a frequency of 122 Hz, the coefficient of t in the sine wave equation is

$$\omega = 2\pi f = 767 \text{ rad s}^{-1}$$

Voltage across the resistor, $V_R = IR = 0.8 \times 56 = 44.8$ V.

Voltage across the capacitor, $V_c = IX_c = 0.8 \times 27.80 = 22.2$ V.

The phasor diagram shows these, with V_C lagging $\pi/2$ behind I and V_R.

As a check, calculate their resultant:

$$v = \sqrt{44.8^2 + 22.2^2} = 50 \text{ V}$$

The equation of i requires only its peak value, and the value of ω:

$$I_{PEAK} = 0.8 \times \sqrt{2} = 1.13 \text{ V}$$

$$\Rightarrow \quad i = 1.13 \sin 767t$$

For the equation of v, which lags behind i we need to calculate the phase angle:

$$\phi = \tan^{-1} \frac{V_C}{V_R} = \tan^{-1} \frac{22.2}{44.8} = \tan^{-1} 0.496 = 0.46 \text{ rad}$$

We can now write the equation for the resultant voltage v, given that V peak $= 50 \times \sqrt{2} = 70.7$

$$v = 70.7 \sin (767t - 0.46)$$

Comment

We can use either rms or peak values of voltage and current when calculating Z, since the ratio of voltage to current is the same for both. However, we keep to one or the other and do not attempt to divide an rms voltage by a peak current. When writing out the equation of an alternating current or voltage, we always use the peak value as the coefficient of the sine.

Reference pages: 30, 66.

RLC series resonant circuit

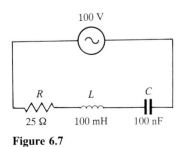

Figure 6.7

Problem

An inductor and capacitor are connected in series (Figure 6.7). The inductor has resistance 25Ω represented by R in the diagram. An alternating voltage of 100V is applied across the circuit. What is the resonant frequency f_o of the circuit? When the circuit is in resonance, what is the current through the circuit and what is the reactance of the capacitor?

$$f_0 = \frac{1}{2\pi \sqrt{LC}} = \frac{1}{2\pi \sqrt{100 \times 10^{-3} \times 100 \times 10^{-9}}}$$

Since $100 = 10^2$, the sum of indices for the expression in brackets is $2 - 3 + 2 - 9 = -8$. Square root this by dividing the index by 2, giving 10^{-4} in the denominator. Get the reciprocal by taking the negative of the index, given 10^4 in the numerator.

Resonant frequency, $f_0 = \dfrac{10^4}{2\pi} = 1592$ Hz

At resonance, $I_0 = V/R = 100/25 = 4$ A

At resonance, the reactance of the capacitor is:

$$X_c = \dfrac{1}{2\pi f C} = \dfrac{1}{2\pi \times 1592 \times 100 \times 10^{-9}} = 1000\ \Omega$$

Reference page: 24.

Phase angle in RLC series circuit

Problem

A 10 mH inductor with coil resistance 25 Ω is connected in series with a 4.7 μF capacitor, wired as in the previous problem. An alternating 100 V signal, frequency 1.2 kHz is applied to the circuit. What is the impedance of the circuit at this frequency? Calculate the current and the voltage phase angle.

Solution

Since we are given the frequency (1.2 kHz = 1200 Hz) we can calculate the reactances of the inductor and capacitor:

$$X_L = 2\pi f L = 2\pi \times 1200 \times 10 \times 10^{-3} = 75.4\ \Omega$$

$$X_C = \dfrac{1}{2\pi f C} = \dfrac{1}{2\pi \times 1200 \times 4.7 \times 10^{-6}} = 28.2\ \Omega$$

The reactances act in opposite directions (Figure 6.8), so the *difference* is used to calculate the impedance of the whole circuit:

$$Z = \sqrt{25^2 + (75.4 - 28.2)^2} = 53.4\ \Omega$$

Since X_L is larger than X_C this difference acts in the direction of X_L.
Having found Z, we calculate the current:

$$I = \dfrac{V}{Z} = \dfrac{100}{53.4} = 1.87\ A$$

The phasor diagram shows Z as the resultant of R and $(X_L - X_C)$. From the diagram, the phase angle is:

$$\varphi = \tan^{-1}\dfrac{47.2}{25} = \tan^{-1} 1.888 = 1.08\ \text{rad} \quad (62.1°)$$

The voltage across the circuit leads the current by 1.08 rad.

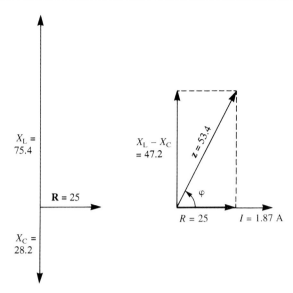

Figure 6.8

Charging a capacitor

Problem

A 100 μF capacitor, which starts by being uncharged, is charged through a 2.2 MΩ series resistor, from a 50 V supply. Plot a graph to show the increase in the voltage across the capacitor until the voltage just exceeds 99% of its final value. From the graph determine (a) how long it takes to charge the capacitor to 43 V, (b) the voltage across the capacitor after 250 s.

Solution

A capacitor charges to within 99% of its final voltage, in $5t$, where t is the time constant. For the circuit given:

$$t = RC = 2.2 \times 10^6 \times 100 \times 10^{-6}$$
$$= 220 \text{s}$$

We need to plot the graph for $t = 0$ to $t = 1100$ s. This is conveniently done for steps of $100t$, using the equation:

$$v_c = V(1 - e^{-t/RC})$$

where v_c is the voltage after time t and V is the supply voltage. The table of values is:

t	0	100	200	300	400	500
V_c	0	18.3	29.9	37.2	41.9	44.8
t	600	700	800	900	1000	1100
V_c	46.7	47.9	48.7	49.2	49.5	49.7

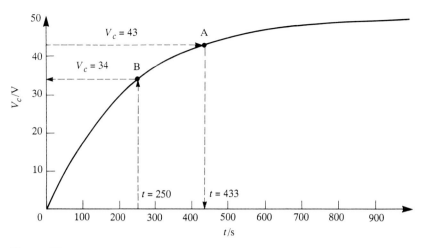

Figure 6.9

The graph is an exponential curve (Figure 6.9) and the answers to the questions are:

(a) Point A is where the line for $V_c = 43$ cuts the curve; reading down to the x-axis, this is the point (43,433). The voltage reaches 43 V after 433 s.
(b) Point B is where the line for $t = 250$ cuts the curve; reading across to the y-axis, this is the point (34,250). The voltage at 250 s is 34 V.

Comments

The result of question (b) could have been calculated, as for the other points on the graph but we treated this as an exercise on graphs. The result for question (a) could be calculated, after changing the subject of the equation to t:

$$t = -RC \ln (1 - V_c/V)$$

Reference pages: 5, 24, 59

Part Two – Maths Topics

This part of the book discusses many different aspects of maths that you may need in your study of electronics. Most of the topics are self-contained and you can turn to them as you need them, in any order. Some chapters require you to have read certain earlier chapters in Parts One or Two. Each chapter begins by listing which earlier topics are required.

7 Distance and direction

> Polar coordinates have several applications in electronics, from radar to ac theory. This chapter explains what they are and how to work with them.
> You need to know about vectors (pages 28–32) and about coordinates of graphs (pages 50–51).

Types of coordinate

The usual way of defining a point on a graph is to state its coordinates. We quote two numbers, the *x*-coordinate and the *y*-coordinate, and generally write them in brackets. These coordinates relate the position of the point to the origin of the graph. Thus a point P(3,7) is three units to the right of the origin and 7 units above it. In this context we take 'above it' to mean 'toward the top of the page'.

Although rectangular coordinates are so familiar to us because of their wide range of uses, they are not the only way in which the position of a point may be specified. Another system makes use of polar coordinates.

Polar coordinates

A sailor wanting to navigate to a small island may locate the island on the map by using its rectangular coordinates, its longitude and latitude, but these are not of practical use in actually sailing to the island. The sailor needs to know what course to steer:

- in *what direction* to point his boat, and
- *how far* to sail.

This is the basis of the system of polar coordinates.

When we use polar coordinates, we define positions with reference to a single point, the **pole**, or **origin** (Figure 7.1). Then all we need to know are the *distance* of the point from the origin and its *bearing*. The bearing is taken with reference to a line from the origin, running horizontally to the right. This is equivalent to the positive *x*-axis on a piece of graph paper. Bearings are expressed in degrees or radians. We write polar coordinates in brackets, the distance followed by the bearing. For example, M in Figure 7.1 is written:

M(3, 56°)

We often use the symbol r for the distance and θ for the angle so that the general symbol for a polar coordinate is:

(r, θ)

Rectangular coordinates can have any value, positive or negative, but polar coordinates are more restricted:

- r is positive
- θ is usually less than 360°

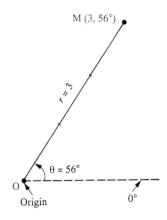

Figure 7.1

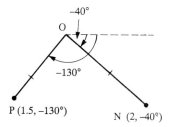

Figure 7.2

The distance of the point is measured from the origin to the point and therefore a negative distance has no meaning. Since 360° is a complete rotation, it makes no sense to have an angle of 360° or more. A point such as (5,361°), for example, is exactly the same point as (5, 1°). Usually we measure θ in the anticlockwise direction, as shown in Figure 7.1, and all values of θ are positive. Sometimes it is more convenient to measure θ in a clockwise direction, especially for points in the third and fourth quadrants (Figure 7.2). If we do this, the angle has a negative value. The units of the angle should be quoted (° or rad) but, when the angle is expressed in terms of π, it is always understood that the unit is the radian.

If the point is actually at the origin, r is zero. In this event we cannot say what value the angle has.

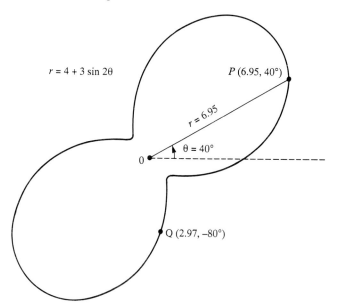

Figure 7.3

We can plot graphs using polar coordinates, either by setting out the angles and distances concerned or by using polar graph paper. Figure 7.3 is an example of a polar graph; it represents the equation:

$$r = 4 + 3 \sin 2\theta$$

Polar graphs may be found in industrial equipment in the form of circular charts used on certain types of recording thermometer and other instruments.

Because of the way radar works, with a steadily revolving array of transmitter and receiver antennas, the information obtained about the positions of aeroplanes, ships or other objects is essentially in polar form—distance and bearing. The raster of the Plan Position Indicator (PPI) display tube is a rotating radius, plotting the polar data in synchrony with the rotation of the antennas.

Explore this

Use a ruler and protractor to plot some polar graphs on plain paper. Here are some equations to get you started:

$r = 3(1 + \cos\theta)$

$r = 1 + \theta/360$ $\quad [0° < \theta < 1440°]$

$r = 2 \sin 4\theta$

Circular motion

Using rectangular coordinates to describe circular motion leads to complicated equations. It is rather like drawing a circle by calculating points on the curve $y^2 = 49 - x^2$ (page 69) and then joining up the points. How much easier it is to use a pair of compasses set to radius 7! Polar coordinates are the equivalent of drawing with compasses.

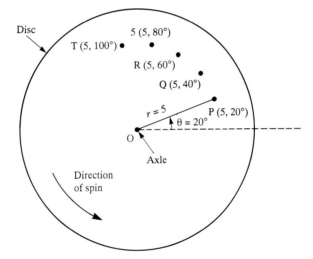

Figure 7.4

Imagine a point P on a rotating disc, which spins about an axle at O (Figure 7.4). If O is also the origin of the coordinate system, the successive positions of P all have the same value of r. For example, if P is at (5,20°) at a given instant of time, its position at successively later times might be:

Q(5, 40°)
R(5, 60°)
S(5, 80°)
T(5, 100°) ... and so on.

Only the angle changes. The rate of change of angle is known as the **angular velocity**. If the angular velocity is constant, the angle increases by an equal amount for equal intervals of time, as in the example above. Angular velocity can be measured in *degrees per second* but is more often expressed in *radians per second*. Its symbol is ω.

Example
A wheel rotates at a rate of 100 revolutions per second. In one second, a radius of the wheel is turned through 36 000°. The angular velocity is:

ω = 36 000 °s^{-1}

Since 360° is equivalent to 2π rad, we can also say that:

ω = 200π rad s^{-1}

Vectors

Vectors may be defined by writing the rectangular coordinates of their starting and finishing points (page 28). Since *direction* and *magnitude* are the two essential features of polar coordinates, just as they are of vectors, there is much to be said for writing vectors in polar form.

When expressing a vector in polar form, we assume that the vector has its starting point at the origin. Then the vector is defined by writing the polar coordinates of its finishing point. The size (r) and direction (θ) of the vector may be written in brackets, as above. In many electronics books, the polar coordinates of vectors are written like this:

$r \underline{|\theta}$

For example, a vector of magnitude 7 and direction 55° is written:

$7 \underline{|55°}$

If the angle is negative, it can be written with a negating sign. Another way to represent a negative angle is to write it as a positive quantity within an inverted sign. For example, we can write $34 \underline{|-78°}$ as:

$34 \overline{|78°}$

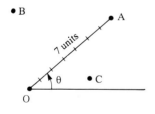

Figure 7.5

Test yourself 7.1

1 Write the polar coordinates of the points A to E in Figure 7.5. Point A shows the scale on which the figure is drawn.

2 On a suitably scaled diagram, plot the points F to J represented by the following polar coordinates:

   a F(3, 50°)   **b** G(7, –20°)
   c H(5, π/2)   **d** I(4, –120°)   **e** J(2, 170°)

3 A wheel is rotating at constant angular velocity, ω. At $t = 0$ s a point P on the wheel has coordinate P(3,10°), the origin of the coordinate being on the axle of the wheel (see Figure 7.4). At $t = 0.1$ s, the point is at P(3,20°). What will be the coordinate of P when:

   a $t = 0.5$ s

b $t = 3.6\,\text{s}$

c Calculate the angular velocity of the wheel, assuming that it turns less than a whole revolution in 0.1 s.

4 The angular velocity of a wheel is $\omega = 157\,\text{rad s}^{-1}$. How many revolutions does it make in 1 s?

5 Assuming that the points A–E in Figure 7.5 represent the finishing points of vectors, all with starting points as O, write the vectors in the form $r\,\lfloor\theta$.

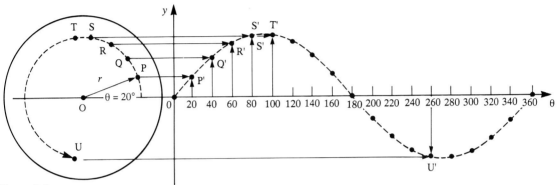

Figure 7.6

Circular motion, vectors and sine waves

Figure 7.6 shows the rotating wheel spinning, as before, with one of its radii marked OP. As the wheel rotates, the radius rotates about the origin O. We can think of this as a vector of constant magnitude (r) with its direction changing steadily because of the circular motion of the wheel. When timing began (when $t = 0$) the vector was lying along the x-axis but, at the instant shown in the figure, it has turned 20°C. At that instant, the finishing point of the vector is at point P, a certain distance above the axis. This distance is shown on a graph drawn to the right of the wheel. The points P and P' are equal distances above the axis. The points on this graph show the heights of the end of the vector above the axis at successive instants in time. Points P', Q', R', S', T', ... , U', ... correspond to the tip of the vector as it passes points P, Q, R, S, T, ... U, ... eventually completing one rotation. To put it another way, each of the upwardly-pointing arrows on the graph represents the y-component of the vector at equal intervals during one rotation.

The graph has a familiar shape. It is clear from the fact that the y-component of the vector is $r\sin\theta$ (see page 65) that the graph represents the equation:

$$y = r\sin\theta$$

Circular motion has generated a sine wave. This is the curve that we often see what we examine with an oscilloscope the mains voltage or the voltage from a signal generator (switched to sine waves). The equation of the voltage sine wave has exactly the same form as that in Figure 7.5, though the variables are different. If we substitute:

voltage, v for y
amplitude, A for r and
ωt for θ

we obtain:

$v = A \sin \omega t$

This is the general equation for a voltage sine wave, or **sinusoidal voltage** as it is sometimes called. Substituting ωt for θ is justified because, if the vector rotates at ω radians per second, then the angle turned in t seconds is ωt radians in t seconds.

We have just shown that a rotating vector can be represented by a sine wave. Conversely, a sine wave can be represented by a rotating vector. And, since polar coordinates are ideal for specifying the magnitude and direction of a vector, they are also good for specifying sine waves.

From sine waves to vectors

Figure 7.7 shows a sine wave, with the equation $v_P = A_P \sin \omega t$, and its corresponding vector **P**. We have drawn the vector at the instant when $t = 0$, assuming $\theta = 0°$ at that time.

There is no need to actually *draw* the stages in the vector's rotation. Calculating v at any given instant is simply a matter of using the wave equation. What we have done is to show by the position of the vector that $v_P = 0$ when $t = 0$. In other words, the direction in which the vector is drawn establishes the initial value of v_P.

Figure 7.7 also shows a second alternating voltage v_Q, amplitude A_Q. It alternates at *the same rate* as v_P, that is, it takes as long to go through one cycle. But it reaches its maximum voltage a short time *after* the first wave reaches its maximum. At all stages, v_Q lags behind v_P. We say that it is **out of**

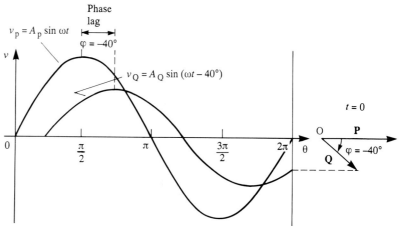

Figure 7.7

phase. When we begin timing at $t = 0$, the value of v_Q has still to rise a little way toward zero. We represent this voltage by another vector **Q**, drawn so that its direction is a little way behind (clockwise of) **P**. The equation of this sine wave is:

$$v_Q = A_Q \sin(\omega t - 40°)$$

This equation includes the phase lag, which is subtracted from ωt when we plot the curve. The corresponding vector **Q** is drawn with magnitude A_Q and at an angle 40° behind **P**. Once again we have drawn the vector in the position it has when $t = 0$. The diagram of the two vectors shows not only their relative magnitudes but also the extent to which they are out of phase. The more they are out of phase, the larger the angle between them. This angle is known as the **phase angle**, φ.

When interpreting this diagram it is important to remember that the two vectors must rotate at the same rate. Otherwise the angle between them would change. Think of them as rotating steadily, but always keeping 40° apart. In terms of sine waves, the two waves take equal times to complete one cycle. In other words, they have the *same frequency*. Diagrams of this kind can be used only when two (or more) signals have equal frequency.

Vectors used to represent the sizes and phases of alternating signals are generally referred to as **phasors**. In a circuit which has an alternating supply voltage there may be many different voltages across the various components of the circuit. The voltages are alternating *at the same frequency*, but they are often of different amplitudes and are usually out of phase with each other. For example, the alternating voltage across a capacitor in the circuit may be widely out of phase with the voltage across an inductor. Drawing a phasor diagram is a powerful way of analysing the relationships between these alternating quantities. We shall see examples of this later in the book.

It might seem from the description above that phasors apply only to quantities which vary according to a simple sine equation. This is true, but we shall see how later that *any* periodic signal, even a pulsed square wave, can be thought of as the sum of a number of sine waves of different amplitudes and frequencies. Each of these component waves can be represented by a phasor.

Test yourself 7.2

1 Given the three sine waves of Figure 7.8, write the equation for y_P, y_Q and y_R. Draw a phasor diagram to show the corresponding phasors, **P** (already drawn), **Q** and **R**, with their relative magnitudes and directions.

2 Given the three phasors of Figure 7.9, write the equations of the sine waves they represent (take $\theta = \omega t$).

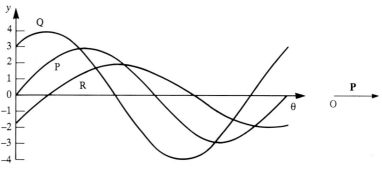

Figure 7.8

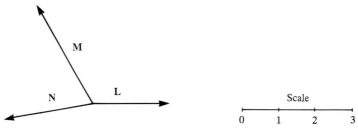

Figure 7.9

Rectangular to polar

Quite often we are given coordinates in one system but need to convert them to the other system to make the calculation easier. Figure 7.10 shows how to convert rectangular coordinates to the polar form, using simple geometry.

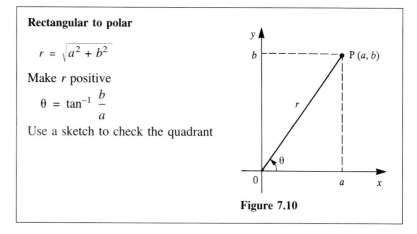

Rectangular to polar

$r = \sqrt{a^2 + b^2}$

Make r positive

$\theta = \tan^{-1} \dfrac{b}{a}$

Use a sketch to check the quadrant

Figure 7.10

In Figure 7.11 the distance r is found by using Pythagoras' theorem (we take the *positive* square root), and θ is found by using the tangent formula. For example, the polar coordinates of point A are:

$r = \sqrt{6^2 + 9^2} = \sqrt{117} = 10.8$ (3 sf)

$\theta = \tan^{-1}(9/6) = \tan^{-1} 1.5 = 56.3°$ (3 sf)

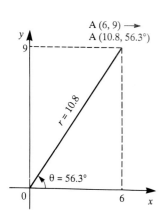

Figure 7.11

In polar coordinates, the point is:

A(10.8, 56.3°)

There is one complication here, because there are *two* angles between 0° and 360° with the tangent 1.5 The other angle is 236.3° in the third quadrant (see box, page 68). In our example, we have a diagram which makes it clear that the angle is not 236.3°. If possible, you should always draw a sketch to check which quadrant the point is in.

Polar to rectangular

We use the sine and cosine formulas for this conversion, as shown in the box. For example, in Figure 7.12, the rectangular coordinates of point B are:

$a = 6 \cos 48° = 4.01$ (3 sf)

$b = 6 \sin 48° = 4.46$ (3 sf)

In rectangular coordinates, the point is:

B(4,01, 4.46)

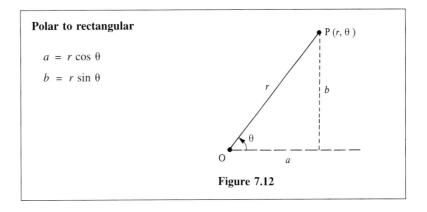

Polar to rectangular

$a = r \cos \theta$

$b = r \sin \theta$

Figure 7.12

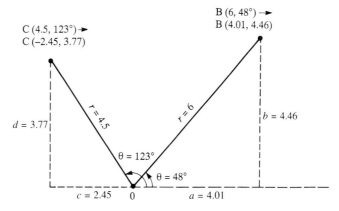

Figure 7.13

If the point is not in the first quadrant, either the sine, the cosine or both are negative, giving the corresponding rectangular coordinates the correct signs. For example, the corresponding coordinates of point C are:

$c = 4.5 \cos 123° = 4.5 \times -0.5466 = -2.45$ (3 sf)

$d = 4.5 \sin 123° = 4.5 \times 0.8387 = 3.77$ (3 sf)

> **Test yourself 7.3**
>
> **1** Convert these rectangular coordinates into the polar form (1 dp).
> [Hint: draw a sketch to check the angle.]
>   **a** A(4,7)   **b** B(−3,6)
>   **c** C(−9, −2)   **d** D(7, −1)
>
> **2** Convert these polar coordinates into rectangular form (1 dp).
>   **a** A(30, 22°)   **b** B(12, 131°)
>   **c** C(15, π)   **d** D(46, −31°)

8 Limits

> The topic of limits is essential to the understanding of differentiation and integration, and these two branches of maths are important in many fields of practical electronics.
>
> You need to know about factors (page 5), algebraic fractions (page 16), indices (page 23), quadratic equations (page 39), graphs (page 49) and functions (page 72).

The term *limit* is used in several different ways in maths. One meaning of the term has the sense of a *boundary*. We may define the domain (page 72) of a function by stating its upper and lower limits. For example, given the function:

$$f(x) = x^2 + 5 \quad 3 \leq x \leq 7$$

The statement on the right specifies that the lower and upper limits of the domain are 3 and 7 respectively. This straightforward meaning of the term is found in several other contexts.

The way *limit* is used in this chapter is best illustrated by an example. Consider a function of x, such as:

$$f(x) = 2x + 3$$

What does the value of $f(x)$ become as the value of x approaches 1? The graph of the function (Figure 8.1) clearly shows that, as x gets closer and closer to 1, $f(x)$ gets closer and closer to 5. We say that 5 is the **limit** of $f(x)$ as x approaches 1. Putting this sentence into maths symbols, we write:

$$\lim_{x \to 1} f(x) = \lim_{x \to 1} 2x + 3 = 5$$

The letters 'lim' stand for 'the limit of' and the symbols '$x \to 1$' below it stand for 'as x approaches 1'.

We use a graph to illustrate the idea of the limit, but we can find the limit without plotting a graph. We simply substitute $x = 1$ in the equation:

$$\lim_{x \to 1} f(x) = 2 \times 1 + 3 = 5$$

In the same way, by substitution, we can find the limit of the function as x approaches other values. For example:

$$\lim_{x \to 0} f(x) = 2 \times 0 + 3 = 3$$

$$\lim_{x \to 60} f(x) = 2 \times 60 + 3 = 63$$

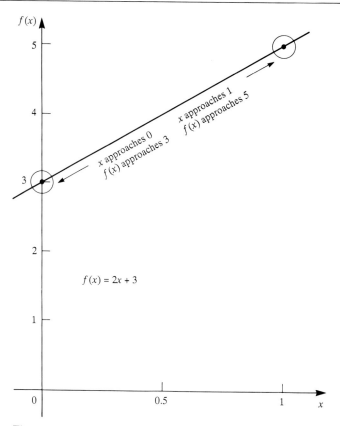

Figure 8.1

Now to take a more complicated function, such as:

$$\lim f(x) = \frac{x+2}{x-3}$$

Finding the limit as x approaches 1 gives the obvious result:

$$\lim_{x \to 1} f(x) = \lim_{x \to 1} \frac{x+2}{x-3} = \frac{3}{-2} = -1.5$$

The graph (Figure 8.2) shows this result at point (1, −1.5)

Finding the limit as x approaches 3 gives a different kind of result:

$$\lim_{x \to 3} f(x) = \lim_{x \to 3} \frac{x+2}{x-3} = \frac{3}{0}$$

The value of 3/0 cannot be calculated because it involves the meaningless operation of dividing by zero. This result is described as *indeterminate*. In other words, there is *no limit* to $f(x)$ as $x \to 3$; it just gets bigger and bigger. This is seen in the graph; the curve falls steeply down and it approaches but never actually reaches the line $x = 3$. $f(x)$ gets more and more negative, and so there is no limit.

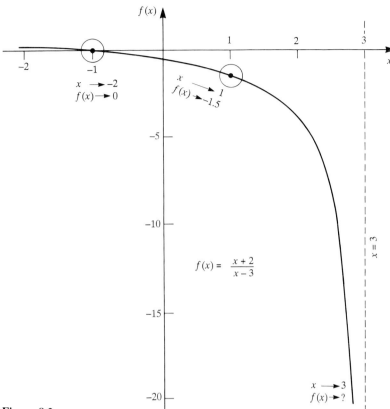

Figure 8.2

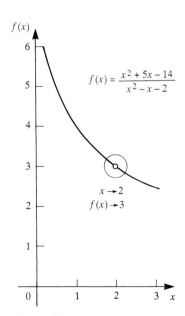

Figure 8.3

Another limit to be thought about is when $x \to -2$.

$$\lim_{x \to -2} f(x) = \lim_{x \to -2} \frac{x+2}{x-3} = \frac{0}{-5} = 0$$

The numerator is zero, so the limit is zero. This is where the curve crosses the x-axis.

In this next example there seems to be no limit as x approaches 2:

$$f(x) = \frac{x^2 + 5x - 14}{x^2 - x - 2}$$

Try to find the limit by substitution:

$$\lim_{x \to 2} f(x) = \lim_{x \to 2} \frac{x^2 + 5x - 14}{x^2 - 2} = \frac{0}{0}$$

Does 0/0 indicate a zero limit? Or is it indeterminate? The graph (Figure 8.3) is plotted with a 'hole' at (2,3) indicating that this function has no value for $x = 2$. Looking at the graph, we can see that the curve approaches the point (2,3) from either direction, even though it does not actually exist at that point. The closer we get to $x = 2$, the closer we get to $y = 3$, so this is obviously the limit.

Although we can follow how a function varies by drawing its graph, there may be other examples in which it is not easy to plot an accurate graph and obtain an accurate value of the limit from it. It is better to find a way of calculating the limit directly, without going to the trouble of plotting a graph. It is often possible to work on the function before trying to find the limit. Let us see what we can do to the function of Figure 8.3. First simplify the function by factorizing and cancelling:

$$f(x) = \frac{x^2 + 5x - 14}{x^2 - x - 2} = \frac{(x + 7)(x - 2)}{(x + 1)(x - 2)} = \frac{x + 7}{x + 1}$$

Having simplified it, find the limit of this new expression:

$$\lim_{x \to 2} f(x) = \lim_{x \to 2} \frac{x + 7}{x + 1} = \frac{9}{3} = 3$$

The simplified function has a limit at 3, confirming what we saw on the graph.

In the last three examples, we found that substituting for x gave a zero in the numerator or in the denominator or in both. The box summarizes what to do in these three cases.

Finding limits

Try substitution.

Result = $0/n$ → limit is zero.

or

Result = $n/0$ → no limit.

or

Result = $0/0$ → simplify and try again.

Test yourself 8.1

Find the limits of the following expressions as x approaches the stated value or, if this is not possible, answer 'No limit'.

1 $5x + 4$, as $x \to 2$

2 $\dfrac{3x - 1}{x}$, as $x \to 1$

3 $\dfrac{x + 18}{x - 4}$, as $x \to 4$

4 $\dfrac{x^2 - 9}{x - 3}$, as $x \to 3$

5 $\dfrac{3 - x}{2 + x}$, as $x \to 2$

6 $\dfrac{2x - 4}{x^2}$, as $x \to 2$

7 $\dfrac{x^2 + x - 2}{x - 1}$, as $x \to 1$

4 $\dfrac{2x^2 - 3x - 9}{x - 3}$, as $x \to 3$

Limits at zero

In Chapters 9 and 10 we shall often need to discover the limits of a function of x as x tends toward zero. As might be expected, just substituting $x = 0$ into the function often leads to one of the cases shown in the box. The methods of finding limits at zero are the same as those we have already looked at.

Limits at infinity

Sometimes we need to be able to predict what happens to a function of x as x tends toward infinity in one direction or the other. Some functions may become infinitely large, others may tend toward zero, while others may settle to some non-zero value. It is not possible to substitute 'infinity' directly into the function as this is bound to give an indeterminate result. We use a special technique.

Examples

Find $\lim_{x \to \infty} \dfrac{3x}{x + 1}$

Look for the highest power of x. In this example, the highest power is x. Divide all terms in the expression by x:

$$\frac{3x/x}{x/x + 1/x} = \frac{3}{1 + 1/x}$$

As x becomes very large, $1/x$ becomes very small. It becomes so small that it can be ignored and the expression becomes:

$$\frac{3}{1 + 0} = 3$$

The limit of the expression is 3.

Find $\lim_{x \to \infty} \dfrac{4x^2 - 3x + 1}{x^2 - 4}$

In this expression x^2 is the highest power, so divide throughout by x^2:

$$\frac{4x^2/x^2 - 3x/x^2 + 1/x^2}{x^2/x^2 - 4/x^2}$$

This simplifies to:

$$\frac{4 - 3/x + 1/x^2}{1 - 4/x^2}$$

When x is very large, both $1/x$ and $1/x^2$ become very small and can be ignored. The expression reduces to:

$$\frac{4 - 0 + 0}{1 - 0} = 4$$

The limit is 4.

> **Test yourself 8.2**
>
> Find the limits of these expressions as x approaches zero.
>
> 1. $\dfrac{x+1}{x-1}$
> 2. -1^x
> 3. $\dfrac{2x^2 + 3x}{x}$
> 4. $\dfrac{x^3 - 3x + 4}{x^2 + 5}$
>
> Find the limits of these expressions as x approaches infinity.
>
> 5. $\dfrac{3x+4}{2x-3}$
> 6. $\dfrac{2x^2 + 3x - 7}{x^2 - 10x + 1}$
> 7. $\dfrac{(2x-3)(5x+2)}{2x^2 + 4}$

Using a calculator

A calculator is a valuable tool for finding limits, especially when:

- the function is complicated,
- factorizing or other simplifying is not possible (or not easy),
- the limit calculation gives a 0/0 result.

This example explains the technique. The function is a rectangular hyperbola (page 69):

$$f(x) = \frac{2x+1}{x-2}$$

With a hyperbola, we expect $f(x)$ to tend toward a limit (one of the asymptotes, page 59) as x becomes infinitely large. We use a calculator (or write a simple program in BASIC for a microcomputer) to work out the value of the function for a series of increasing values of x. Begin with a small, but not too small, value ($x = 10$ is suitable in this example), and finish when the approaching limit becomes obvious.

When investigating limits with x tending toward a large value, it is often a good idea to try successive powers of 10. In the table below the results are given to four decimal places:

x	10	50	100	1000	10 000	100 000	1 000 000
$f(x)$	2.6258	2.1042	2.0510	2.0050	2.0005	2.0001	2.0000

To illustrate the trend in these results, they are graphed in Figure 8.4. The graph is drawn using logs on the x-axis so that all the points can be plotted on one page. But plotting graphs is by no means an essential part of the calculator

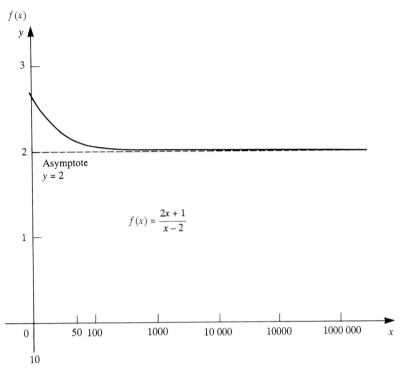

Figure 8.4

method. The table of results alone makes it clear that the value of y is approaching a limit of 2.

The previous example is easily solved by the usual technique for limits at infinity (page 107) and does not need to be solved with a calculator. It is given as an example of the technique and it is readily confirmed that the limit is 2.

In the next example, the function cannot be factorized or simplified in any way:

$$f(x) = (1 + 1/x)^x$$

Find $\lim_{x \to \infty} f(x) = \lim_{x \to \infty} (1 + 1/x)^x$

This equation does not represent any of the well-known kinds of curve. Also, it is difficult to guess how this function varies, just by inspecting the equation. As x ranges up from 1 to a very large value, the part of the function in brackets ranges from 2 down to 1. But at the same time this quantity is being raised to an increasingly large power. Does $f(x)$ become smaller or larger? A calculator supplies the answer (3 dp):

x	1	10	100	1000	10 000	100 000
$f(x)$	2.000	2.594	2.705	2.717	2.718	2.718

As will be explained in a later section, the limit of this function as x becomes infinitely large is the exponential constant, e (page 5).

Test yourself 8.3

Use a calculator to find the limits of these expressions as x approaches the value given or, if this is not possible, answer 'No limit'.

1 $\dfrac{x}{2^x}$, as $x \to 0$ 2 $x^{1/x}$, as $x \to \infty$

3 $\dfrac{\sin x}{x}$, as $x \to \infty$

9 Rates of change

> Most electronic circuits are the scene of rapid changes in electrical quantities. The changing charge on a capacitor is used in timing. Audio amplifiers and filters abound with changing voltages and currents. The rates of change of these quantities is an essential aspect of the way a circuit behaves. The maths of rates of change therefore has very many applications to electronics.
>
> You need to know about graphs (Chapter 5) and limits (Chapter 8).

> **Function notation**
>
> $$y = 2x + 3 \quad \text{and} \quad f(x) = 2x + 3$$
>
> are two ways of defining the same function. The y-notation is shorter. In this chapter we use the y-notation except where we need to specify a particular value of x. Then we use the f-notation.

The graph of the function $y = 2x + 3$ (Figure 8.1) is a straight line, sloping up to the right with gradient $m = 2$. We deduce these facts from the equation alone, without the need to plot the graph, by comparing the function with the general equation of a straight-line graph:

$$y = mx + c$$

In this equation, m is the gradient. The definition of the gradient is an important one, which we shall refer to often in this chapter, so we repeat it here:

$$\text{gradient, } m = \frac{\text{increase in } y}{\text{increase in } x}$$

Since the line in Figure 8.1 is straight, the ratio has the same value for any segment of the line. In other words, the line has the same gradient along the whole of its length.

There are many functions which produce lines that are *not* straight. The gradient varies at different points along the line. As an example, the graph of Figure 9.1 represents the quadratic function:

$$y = x^2 - 4x + 7$$

The gradient is negative when x is less than 2. As x approaches 2 (from lower values) the gradient becomes less and less until, when $x = 2$ exactly, the gradient is zero. From then on, the gradient is positive and increases with increasing values of x. There is no part of the line which is straight, so there is no place where the gradient is constant.

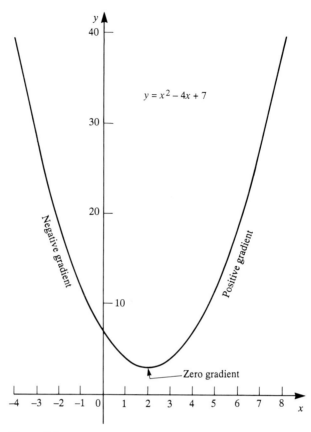

Figure 9.1

Pin-pointing the gradient

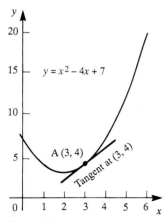

Figure 9.2

Let us investigate the gradients of Figure 9.1 more closely. Suppose we want to know what the gradient is at a particular point; when $x = 3$, for example. The slope of the curve at that point is the same as that of a line which just *touches* the curve at that point. A direct way of measuring the gradient is to place a straight edge against the curve at that point, draw the line which just touches the curve (the **tangent**) and then measure its slope. The tangent is drawn in Figure 9.2, which shows part of the curve plotted on a larger scale. Drawing the tangent is easy to describe but difficult to do. Plotting a sufficiently accurate curve is hard enough. Drawing the tangent at precisely the right point and with precisely the right slope is harder. As we shall show later, a little algebra replaces the drawing and gives an exact result, too.

Before going on to the algebra, we will try another approach, still using the graph. The point at which we want to find the gradient, when $x = 3$, is referred to as point A. By substitution, we calculate that, for point A:

$$y = f(3) = x^2 - 4x + 7$$
$$= 9 - 12 + 7 = 4$$

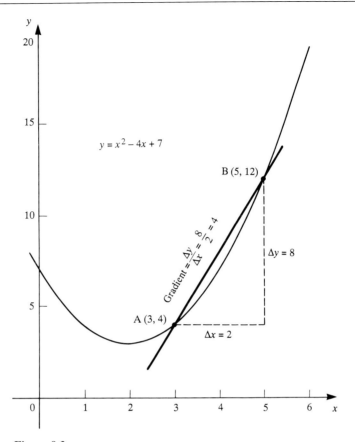

Figure 9.3

The coordinates of A are (3,4). A practical but very rough method for estimating the gradient at A is to draw a straight line *not* touching the curve but *cutting* the curve at A and also at a point B, further along the curve to the right (Figure 9.3). Comparing the line AB with the tangent in Figure 9.2, it is obvious that the line is steeper than the tangent so the gradient of the line is an overestimate of the gradient of the curve at A.

To get from A to B we move a distance to the right and a distance up the page. In Figure 9.3 the distance between A and B to the right is labelled by the symbol Δx. This symbol does *not* mean 'Δ multiplied by x'. The Δ stands for the words 'a small increment in', so Δx means, 'a small increment in x'. In the figure, Δx is a distance of 2 units, which is not particularly small, as it had to be drawn a reasonable size to make the figure clearer. It will be made smaller later.

By substituting $x = 5$ in the equation, we find that the y-coordinate of B is 12. The corresponding increment in y is:

$$\Delta y = f(5) - f(3) = 12 - 4 = 8$$

Summing up: while x increases from 3 to 5, a horizontal distance of 2, y

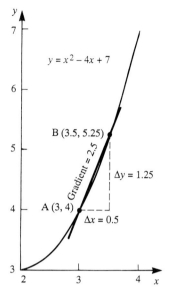

Figure 9.4

increases from 4 to 12, a vertical distance of 8. Calculating m for the line AB gives:

$$\text{gradient, } m = \frac{\text{increase in } y}{\text{increase in } x} = \frac{\Delta y}{\Delta x} = \frac{8}{2} = 4$$

The gradient of the line is 4. This estimate is in error simply because the straight line is not the curved line. The gradient of the curved line at A can be seen to be less than that of the straight line. At B, the gradient of the curve is greater than that of the line. But 4 is the best estimate we have available so far.

A better estimate can be made if we move B closer to A. Then the curve between the two points does not bulge out as much. It is closer to being a straight line. We get a more precise estimate of the gradient.

Suppose B is now located a horizontal distance $\Delta x = 0.5$, to the right of A. This is shown in Figure 9.4, in which a small part of the curve is drawn on an even larger scale. If the x-coordinate of B is $3 + 0.5 = 3.5$, the y-coordinate is:

$$f(3.5) = 3.5^2 - 4 \times 3.5 + 7 = 5.25$$

For point B, $\Delta x = 0.5$ and $\Delta y = f(3.5) - f(3) = 5.25 - 4 = 1.25$. The new estimate of the gradient is:

$$m = \frac{\Delta y}{\Delta x} = \frac{1.25}{0.5} = 2.5$$

Even so, the curve is not *quite* the same as the straight line so the estimate remains an estimate, not an exact result.

To improve precision still further, we could move B even closer to A. This process of moving B closer and closer to A recalls the idea of finding limits, described in the previous chapter. As the distance between A and B becomes smaller and smaller, we obtain a better and better estimate of the gradient. The problem of finding the gradient at A thus reduces to one of calculating the *limiting value* of m as the distance between A and B approaches zero. In geometrical terms, we are finding the limit of the gradient as the two points get so close together that they become the *same* point and the straight line becomes the *tangent*.

In mathematical terms, we need to calculate:

$$\lim_{\Delta x \to 0} m = \lim_{\Delta x \to 0} \frac{\Delta y}{\Delta x}$$

Simplifying the symbols

Before we discover the best way of calculating the limit of the gradient, we introduce a new set of symbols to replace the rather cumbersome collection on the right of the equation above. The new set is:

$$\frac{dy}{dx}$$

RATES OF CHANGE 115

This symbol is pronounced 'dee y by dee x'. Each 'Δ' is replaced by a 'd'. As with Δ, dy does *not* mean 'd *times* x'. It means 'a small increment when the increment is becoming zero'. In other words, dy/dx is Δy/Δx at its limit, when Δx has become zero. So, when using dy/dx, there is no need to precede it with the symbols 'lim' and 'Δx→0' as these are implied.

Finding dy/dx

An obvious way of finding the limit of *m* is to use a calculator, as on page 108, to find successive values of it as B moves closer and closer to A. Eventually we should find *m* approaching a steady value, its limit, the gradient at A. Although this is a practicable technique, it can become tedious when the function is complicated. There are easier ways of arriving at the same answer.

If we look at the problem in general terms instead of getting involved with particular values of *x* and Δx, it becomes much simpler. On page 113 we estimated *m* as B was moved closer to A. At each stage, the calculation was the same. Given that A is point (x,y) and that B is at (x+Δx, y+Δy), this is how we calculate *m*:

$$m = \frac{\Delta y}{\Delta x} = \frac{[y\text{-coordinate of B}] - [y\text{-coordinate of A}]}{\Delta x}$$

$$= \frac{\begin{bmatrix} \text{Value of the function} \\ \text{when we put } (x + \Delta x) \\ \text{into it} \end{bmatrix} - \begin{bmatrix} \text{Value of the function} \\ \text{when we put } x \\ \text{into it} \end{bmatrix}}{\Delta x}$$

We then find the limit of *m* as Δx approaches zero.

As an example, but without deciding on any particular values for *x* or δx, we use the same function on page 111:

$$y = x^2 - 4x + 7$$

Now we put it into the equation above:

$$m = \frac{[(x + \Delta x)^2 - 4(x + \Delta x) + 7] - [x^2 - 4x + 7]}{\Delta x}$$

In the square brackets on the left is the function with (x + Δx) instead of *x*, giving the y-coordinate of B. In the square brackets on the right is the function, giving the y-coordinate of A. This rather long expression can be simplified, remembering to change the signs of terms in the right-hand square brackets:

$$m = \frac{x^2 + 2x\Delta x + (\Delta x)^2 - 4x - 4\Delta x + 7 - x^2 + 4x - 7}{\Delta x}$$

There are several pairs of terms of opposite sign which cancel out, leaving:

$$m = \frac{2x\Delta x + (\Delta x)^2 - 4x\Delta}{\Delta x}$$

Divide the numerator by the denominator, Δx:

$$m = 2x + \Delta x - 4$$

The limit of m is when Δx becomes zero:

$$\lim_{\Delta x \to 0} m = \frac{dy}{dx} = 2x + 0 - 4$$

The gradient at A is therefore given by the function:

$$m = \frac{dy}{dx} = 2x - 4$$

A calculation which was very elaborate in its early stages has provided a very concise result. Soon we shall discover how to avoid the complicated part of the calculation altogether.

Note that we have not used any particular values of x in this calculation. It applies *anywhere* along the curve. To find the gradient at any point on the curve, all we have to do is to substitute the value of x into this equation. For point A, where $x = 3$:

$$m = 2 \times 3 - 4 = 2$$

Using the same equation, we find the gradient at other points, for example:

x	m
-2	-8
0	-4
1	0
5	6
10	16
100	96

Test yourself 9.1

Use the limit method described above to find the gradients of these functions at the points specified.

1. $y = x^2 + 5x - 3$, when $x = 5$
2. $y = x^2 - 11x + 2$, when $x = 0$
3. $p = q^2 + 4q - 9$, when $q = -3$
4. $y = 3x^2 + 2x + 4$, when $x = 1$
5. $y = x^3 + x^2 + x + 10$, when $x = 7$
   [Hint: $(x + \Delta x)^3 = x^3 + 3x^2 \Delta x + 3x(\Delta x)^2 + (\Delta x)^3$.]
6. $g = 4h^3 - 5h^2 + 13h - 6$, when $h = 2$

RATES OF CHANGE

A function from a function

The calculation of the previous section (page 111) began with a function of x:

$$y = x^2 - 4x - 7$$

This tells us how to plot the curve.
The calculation ended with another function of x:

$$\frac{dy}{dx} = 2x - 4$$

This tells us the gradient of the curve at any given point.

The second function is *derived* from the first function, so we refer to dy/dx as the **derived function** of y, or the **derivative** of y. It is also known as the **differential** of y, and the process of calculating it is known as **differentiation**.

Looking for the rules

It would help avoid the rather long (and error-prone) calculations of differentiation if we could find some rules that would allow us to convert a function into its derivative straight away. Below are listed the functions, and derivatives of the earlier example and also the examples from 'Test yourself 8.2'.

Function (y)	*Derivative* (dy/dx, etc.)
$x^2 - 4x + 7$	$2x - 4$
$x^2 + 5x - 3$	$2x + 5$
$x^2 - 11x + 2$	$2x - 11$
$q^2 + 4q - 9$	$2q + 4$
$3x^2 + 2x + 4$	$6x + 2$
$x^3 + x^2 + x + 10$	$3x^2 + 2x + 1$
$4h^3 - 5h^2 + 13h - 6$	$12h - 10h + 13$

By contrasting the two columns of the table, a definite pattern can be seen:

- The number of terms in the derivative is one less than in the function. The last term in each function is a constant which does not contribute to the derivative.
- In the derivatives, the index of x is 1 less than it is in the corresponding term of the function. This includes terms with index 1, such as $5x$ which have index 0 in the derivative. The x does not appear in the derivative since $x^0 = 1$.
- The sign of the term is unchanged in these examples (though it will be shown later that it does change if the index of x is negative).
- The coefficient of each term in the derivative is its original coefficient multiplied by the index of x in the original. For example:

$$x^2 \Rightarrow 2 \times x = 2x$$
$$3x^2 \Rightarrow 2 \times 3x = 6x$$
$$x^3 \Rightarrow 3 \times x^2 = 3x^2$$
$$4h^3 \Rightarrow 3 \times 4h^2 = 12h^2$$

In these examples the index has also been reduced by 1, in accordance with the rule already stated above.

The rules are summarized in the box. We have deduced the rules by looking at typical examples of differentiation. You can see that they are true, for these examples at least. The rules can be proved mathematically to be true for all similar functions.

Rules for differentiating a simple function

Multiply the term by the index of x.
Reduce the index by 1.

Summarizing:

$$y = x^n \rightarrow dy/dx = n \cdot x^{n-1}$$

or, if the term has a coefficient, a

$$y = ax^n \rightarrow dy/dx = na \cdot x^{n-1}$$

The derivative of a constant term is zero.

Other symbols

The most common symbol for a derivative is:

$$\frac{dy}{dx}$$

This is the one used almost always in this book and in many other books. The variables may be different, according to the context, but all indicate the process of differentiation. Examples include:

$$\frac{du}{dx} \qquad \frac{di}{dt} \qquad \frac{dv}{dt}$$

On some occasions it may save time and space to symbolize the derivative directly, without first defining the function y. The usual way is:

$$y = x^2 + 3 \Rightarrow \frac{dy}{dx} = 2x$$

The more direct way is to write $\frac{d}{dx}$, followed by the function to be differentiated, in brackets:

$$\frac{d}{dx}(x^2 + 3) = 2x$$

Derivatives are also written using the function notation (page 72). Here are the equivalents of the dy/dx forms:

The function:

$$y = 3x^2 + 4x + 5 \qquad \Rightarrow \qquad f(x) = 3x^2 + 4x + 5$$

First derivative:

$$\frac{dy}{dx} = 6x + 4 \qquad \Rightarrow \qquad f'(x) = 6x + 4$$

Second derivative (page 124):

$$\frac{d^2y}{dx^2} = 6 \qquad \Rightarrow \qquad f''(x) = 6$$

Using the rules

The method used on page 115 for finding a derivative by calculating its limit is known as **differentiating from first principles**. The technique is a basic one which illustrates what differentiation is about, but it is a lengthy one. Given the rules for differentiating, as set out in the box, there is no need to go back to first principles. Here are some examples of functions easily differentiated on sight by using the rules:

y	dy/dx
$3x^3$	$9x^2$
$-5x^4$	$-20x^3$
$2x^7$	$14x^6$
$2x^4 + 3x^2$	$8x^3 + 6x$
$x^5 - 2x^3 - 3$	$5x^4 - 6x^2$

The rules also apply to negative and fractional powers of x. An example of x having a negative power is:

$$y = \frac{1}{x} = x^{-1}$$

According to the rules, the expression is multiplied by its index, -1. Then the index is reduced from -1 to -2:

$$\frac{dy}{dx} = -1 \cdot x^{-2} = \frac{-1}{x^2}$$

Here is an example of a function in which x has a fractional power:

$$y = \sqrt{x} = x^{1/2}$$

The expression is multiplied by its index, ½. Then the index is reduced from ½ to −½:

$$dy/dx = \tfrac{1}{2} \cdot x^{-1/2} = \frac{1}{2\sqrt{x}}$$

Rates of change in electronics

Rates of change are an essential feature of many electronic circuits. Current is a rate of flow of electric charge. Given constant current I, the charge q on a capacitor (starting from zero charge) at time t is:

$$q = It$$

A small q indicates **instantaneous charge**, the charge at any given instant of time, t. The voltage across the capacitor (Figure 9.5) is:

$$v = \frac{q}{C} = \frac{I}{C} \cdot t$$

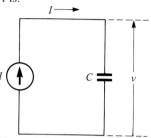

Figure 9.5

Figure 9.6 is an example of a graph of this function. The rate at which the voltage is changing is the gradient:

$$\frac{dv}{dt} = \frac{I}{C}$$

For example, given current 250 µA and capacitance 100 µF:

$$\frac{dv}{dt} = \frac{250 \times 10^{-6}}{100 \times 10^{-6}} = 2.5$$

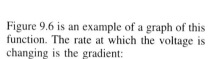

Figure 9.6

The rate of change of voltage is 2.5 volts per second. Note that t does not appear in the derivative, which means that the rate of change of voltage is the same *at all times*. It is constant, at least for as long as we can hold the current constant. Its graph is a straight line.

Opposing change

A changing current induces an emf in the coil (Figure 9.7) proportional to the rate of change of current. If L is the self-inductance of the coil, the emf e is given by:

$$e = -L \cdot \frac{di}{dt}$$

In this equation di/dt is the rate of change of current. The negative sign indicates that the emf *opposes* the change of current (back emf). For example, if V is varying so that the function for the instantaneous current is:

$$i = 3t^2 - 4$$

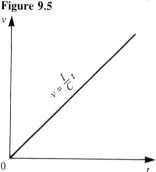

Figure 9.7

The rate of change of current is the derivative of this:

$$\frac{di}{dt} = 6t$$

If $L = 10\,\mu H$ and $t = 2\,s$, the opposing emf at that instant is:

$$e = -10 \times 10^{-6} \times 6 \times 2 = -120 \times 10^{-6} = -120\,\mu V$$

A circuit for differentiating

The operational amplifier (Figure 9.8) is wired as a differentiator. The relationship between input and output voltage is given by:

$$v_{OUT} = -RC \cdot \frac{dv_{IN}}{dt}$$

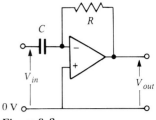

Figure 9.8

- v_{OUT} is negative if v_{IN} is rising ($dv/dt > 0$), and the faster it is rising the lower v_{OUT} falls.
- v_{OUT} is zero if there is no change in input voltage ($dv/dt = 0$).
- v_{OUT} is positive if v_{IN} is falling ($dv/dt < 0$).

For example, $R = 10\,k\Omega$, $C = 2\,\mu F$, and v rises 2 V in 4 s, then $dv/dt = 0.5$ and:

$$v_{OUT} = -10 \times 10^3 \times 2 \times 10^{-6} \times 0.5 = 0.01\,V$$

Test yourself 9.2

Use the rule $y = ax^n \Rightarrow dy/dx = n \cdot ax^{n-1}$ to find the derivatives of these functions.

1. x^4
2. 7
3. $35x^{21}$
4. $3 - 8x$
5. πx^3
6. $4x^3 - 2x$
7. $1.3x^4$
8. x^p
9. kx
10. $2/x$
11. $4/x^2$
12. $2\sqrt{x}$
13. $4x - 1/x$
14. $4x^2 + x - 2/x^4$
15. $\sqrt[3]{x}$

16. The current i through a 0.5 H inductor (a) rises at a constant rate from zero to 10 mA in 4 s. It then (b) remains constant at 10 mA for 3 s. Finally (c) it falls from 10 mA to zero in 1 s. Calculate the voltage across the inductor at stages (a), (b) and (c).

17. The current through a 100 mH inductor is given by the function: $i = 0.1 - t^2$. Calculate the voltage across the inductor (a) when $t = 0.1$ s, (b) when $t = 0.2$ s, and (c) at the instant when $i = 0$, to 3 sf.

Useful derivatives

By differentiating from first principles, it is possible to obtain derivatives of a wide range of functions. The box summarizes the ones which are most often used in electronics. Of these, the sine and cosine functions are particularly important. This is because alternating currents of all kinds can be analysed in terms of one or more sine waves (page 268).

Functions and derivatives

Function	Derivative	Example	
		Function	Derivative
x^n	$n \cdot x^{n-1}$	x^3	$3x^2$
ax^n	$a \cdot n \cdot x^{n-1}$	$4x^3$	$12x^2$
$\sin a\theta$	$a \cos a\theta$	$\sin 3\theta$	$3 \cos 3\theta$
$\cos a\theta$	$-a \sin a\theta$	$\cos 2\theta$	$-2 \sin 2\theta$
e^{ax}	$a\, e^{ax}$	e^{3x}	$3e^{3x}$
$\ln ax$	$1/x$	$\ln 5x$	$1/x$
a^x	$a^x (\ln a)$	2^x	$2^x (\ln 2)$

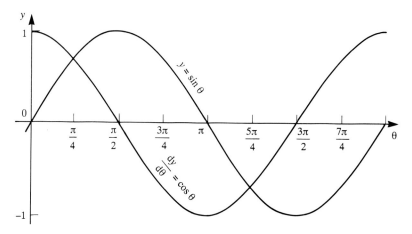

Figure 9.9

Figure 9.9 shows the functions $y = \sin \theta$ and $dy/dx = \cos \theta$ plotted on the same axes. Because the derivative of $\sin \theta$ is $\cos \theta$, the cosine curve shows how the gradient of the sine curve varies with θ. This can be confirmed by examining the curves at a few values of θ:

- When θ is zero, the sine curve is sloping up most steeply to the right. This is its maximum positive gradient. At this point, $\cos \theta$ has its maximum value, +1.
- When θ is $\pi/2$ (90°), the sine curve reaches its maximum. Its gradient is zero. At this point, $\cos \theta$ is zero.
- When θ is π (180°), the sine curve is sloping down most steeply. This is its greatest *negative* gradient. At this point, $\cos \theta$ has its minimum value, −1.

Similar correspondences between the function $\sin \theta$ and its derivative can be seen at other points on their curves.

Test yourself 9.3

Find the derivatives of these functions of x or θ.

1. $y = 2x^{10}$
2. $y = \dfrac{4}{3x^9}$
3. $\sin \theta$, where $\theta = 36°$
4. $2 \cos \theta$, where $\theta = \pi$
5. e^x, where $x = 3$
6. $4 \cdot \ln x$, where $x = 4$
7. 3^x
8. $\ln (7x)$
9. $\sin 4\theta$
10. e^{3x}
11. $4 \cos 2\theta$
12. $3 \sin 2\theta - 2 \cos (\theta/2)$

13. The instantaneous voltage across a 47 μF capacitor is $v = 10 \sin 2t$. Given that the instantaneous current entering the capacitor is $i = C \cdot dv/dt$, calculate the current when $t = 0.1$ s (3 sf).

14. An alternating current has the equation $i = 0.05 \sin 2\pi ft$, where f is the frequency.
    a What is the rate of change of the current when $t = 25$ ms, with frequency 100 Hz? (to 3 sf).
    b What voltage would be produced by this current across an inductor of 250 μH when $t = 0.1$ ms and the frequency is 1 kHz? (to 3 sf).

15. A coil has self-inductance $L = 0.5$ H, and a resistance $R = 60\,\Omega$. A constant voltage $E = 10$ V is applied to it for a few seconds and is then switched off. The instantaneous current through the coil is given by:
    $$i = \dfrac{E}{R}(1 - e^{-Rt/L})$$
    Calculate the rate at which the current changes:
    a when the switch is first opened ($t = 0$), and
    b 25 ms later (3 sf).

16. A 470 μF capacitor is charged to 12 V, then discharged through a 1 MΩ resistor. Given that the instantaneous voltage across the capacitor is:
    $$v = V_0 \cdot e^{-t/RC}$$
    where V_0 is the initial voltage, calculate the rate of change of voltage:
    a as discharge begins ($t = 0$), and
    b 0.2 s later (3 sf).

Maximums and minimums

When a function reaches a maximum or minimum its gradient is zero (page 70). It follows that the derivative of the function is zero at these points. Take, for example, the function graphed in Figure 9.1, page 112:

$$y = x^2 - 4x + 7$$

$$\Rightarrow \quad \frac{dy}{dx} = 2x - 4$$

Putting the derivative equal to zero:

$$2x - 4 = 0$$
$$\Rightarrow \quad x = 2$$

The gradient is zero when $x = 2$. The fact that $dy/dx = 0$ tells us that this is a stationary point (page 70), but is it a maximum, a minimum, or a point of inflection? By looking at the graph we can see that this point is a minimum but, without the graph, we could not be certain about this.

The derivative identifies the stationary points on a curve, but does not say what kinds of stationary points they are.

Figure 5.27 shows the types of stationary point. These are distinguished by what happens to the gradient as the curve passes *through* the stationary point:

- maximum – gradient is decreasing through zero
- minimum – gradient is increasing through zero
- point of inflection – gradient is zero at the point but has the same sign before and after the point

The way to discover the change of gradient (without plotting the graph) is to find the rate of change of the *gradient*. This is, in effect, the *gradient of the gradient*. Or it could be called the 'derivative of the derivative'. We differentiate the function again. These are the stages:

Function	$y = x^2 - 4x + 7$
Derivative	$y = 2x - 4$
Derivative of derivative	$y = 2$

The derivative is zero when $x = 2$, showing that there is a point of inflection there. The derivative of the derivative is 2, showing that the gradient of the gradient is *positive*. It is *increasing* and therefore the point is a minimum.

The derivative of a derivative is known as a **second derivative** or **second differential**. Its symbol is:

$$\frac{d^2y}{dx^2}$$

This is pronounced 'dee two y by dee x squared'.

As another example, take a cubic curve:

$$y = x^3 - 4x^2 + 4x - 10$$

The first derivative is:

$$\frac{dy}{dx} = 3x^2 - 8x + 4$$

Putting this equal to zero, and solving the quadratic this produces:

$$3x^2 - 8x + 4 = 0$$
$$\Rightarrow (3x - 2)(x - 2) = 0$$
$$\Rightarrow 3x - 2 = 0$$
$$\text{or} \quad x - 2 = 0$$
$$\Rightarrow x = 2/3$$
$$\text{or} \quad x = 2$$

There are stationary points at $x = 2/3$ and $x = 2$
The second derivative is:

$$\frac{d^2y}{dx^2} = 6x - 8$$

If $x = 2/3$, the value of the second derivative is $\frac{6 \times 2}{3} - 8 = -4$. This is a negative value, so the point is a maximum.

If $x = 2$, the value of the second derivative is $6 \times 2 - 8 = 4$. This is a positive value, so the point is a minimum.

Although the second derivative is the best way of deciding whether a stationary point is a maximum or minimum, there are cases when this method presents difficulties. It may happen that the first derivative is not easily differentiated. The best method then is to examine the *first* derivative and discover how its value changes if x is made a little smaller and a little larger.

For example, take the function:

$$y = x^3 - \frac{x^2}{2} - 2x + 5$$

$$\frac{dy}{dx} = 3x^2 - x - 2$$

Put this equal to zero to find stationary points:

$$3x^2 - x - 2 = 0$$

This is solved by factorizing:

$$(3x + 2)(x - 1) = 0$$
$$\Rightarrow (3x + 2) = 0 \quad \text{or} \quad (x - 1) = 0$$
$$\Rightarrow x = -3/2 \quad \text{or} \quad x = 1$$

If $x = -2/3$, then
$$y = (-2/3)^3 - \frac{(-2/3)^2}{2} - 2 \cdot -2/3 + 5 = 5.8$$

If $x = 1$, then
$$y = (1)^3 - \frac{(1)^2}{2} - 2 \cdot 1 + 5 = 3.5$$

The stationary points are (–2/3, 5.8) and (1, 3.5).

First and second differentials in action

One type of state-variable filter (shown in block form in Figure 9.10) includes two amplifiers which are differentiators (Figure 9.8). If the input to the first differentiator is a sine wave, $y = \sin\theta$, the output is:

$$\frac{dy}{d\theta} = \cos\theta$$

As a result of differentiation, the output lags the input by $\pi/2$ (Figure 9.9). In the state-variable filter this output is fed to a second differentiating amplifier and it is differentiated again:

$$\frac{d^2y}{dx^2} = -\sin\theta$$

The second derivative is the inverse of the original function. In other words, the output is 180° out of phase with the input to the first differentiator. This is then inverted and reduced in amplitude by the inverting adder and sent back to the input of the first differentiator, acting as positive feedback. Part of the $\cos\theta$ signal is inverted and fed back to control the damping of the filter.

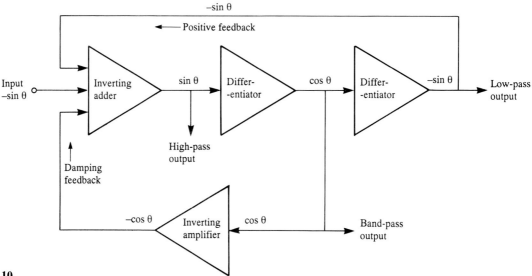

Figure 9.10

Consider the point (–2/3, 5.8). If x is made a little less than –2/3, the expression $(3x + 2)$ becomes negative, $(x - 1)$ remains negative and the derivative is therefore positive (a negative times a negative). If x is made a little more than –2/3, $(3x + 2)$ becomes positive, $(x - 1)$ remains negative and the derivative is negative. Thus the derivative changes from positive to negative at this point. The point (–2/3) is a maximum.

Consider the point (1, 3.5). If x is made a little less than 1, the expression $(3x + 2)$ remains positive, $(x - 1)$ becomes negative and the derivative is therefore negative. If x is made a little more than 1, $(3x + 2)$ remains positive, $(x - 1)$ becomes positive and the derivative is positive. Thus the derivative changes from negative to positive at this point. The point (1,3) is a minimum.

> **Test yourself 9.4**
>
> Find the coordinates of the stationary points on the curves, and say whether each is a maximum or a minimum.
> 1 $y = x^2 - 4x + 5$
> 2 $y = 5 - (3 - x)^2$
> 3 $y = 2x^3 - 6x^2 - 90x + 5$

> The rest of this chapter deals with extensions of the basic techniques of differentiation. These are for you to refer to when you need them.

The Chain Rule

A function such as $y = 3x + 2$ is a **simple** function. A function such as $y = (3x + 2)^2$ is known as a **composite** function. It is a function within a function. In other words, it consists of *two* functions, executed one after the other:

Stage 1: evaluate $3x + 2$.
Stage 2: square the result of Stage 1.

$3x + 2$ is known as the **inner function**, and the squaring of this as the **outer function**. To find the derivative of a composite function, we cannot simply differentiate the terms in brackets and then square the result, for this is finding the derivative of the inner function, but not of the outer function.

The most direct way of tackling this kind of problem is to expand the function by multiplying it out:

$$y = (3x + 2)^2 = 9x^2 + 12x + 4$$

This converts it into a simple function. Differentiating gives:

$$\frac{dy}{dx} = 18x + 12$$

The technique of expanding the function is sometimes easy to apply, but there are often cases in which it is a lengthy matter or even an impossible task to work in this way. This is when we apply the **Chain Rule**. The first step in using the Chain Rule is to set out the two functions separately. Call the inner function u:

$$u = 3x + 2$$

Express the outer function y in terms of u:

$$y = u^2$$

The first equation relates x to u and the second relates u to y. Both equations are simple functions and are differentiated in the usual way:

$$\frac{du}{dx} = 3 \quad \text{and} \quad \frac{dy}{du} = 2u$$

Linking the equations for du/dx and dy/dx together by the Chain Rule restores the link between x and y:

$$\frac{dy}{dx} = \frac{dy}{du} \cdot \frac{du}{dx} = 2u \cdot 3 = 6u$$

But $u = 3x + 2$, so:

$$\frac{dy}{dx} = 6(3x + 2) = 18x + 12$$

This is the same result as we obtained earlier by expanding and then differentiating.

A good example of a complex function that is much more easily differentiated by using the Chain Rule is:

$$y = \sin 3x$$

Writing the derivative as cos $3x$ ignores the fact that '$3x$' is itself a function of x. The two functions are:

Inner function $u = 3x$
Outer function: $y = \sin u$

Differentiating these, we obtain:

$$\frac{du}{dx} = 3$$

and

$$\frac{dy}{du} = \cos u$$

Combining them according to the Chain Rule gives:

$$\frac{dy}{dx} = 3 \cdot \cos u$$

But $u = 3x$, so:

$$\frac{dy}{dx} = 3 \cdot \cos 3x$$

This leads to the general rule that the derivative of sin ax is a cos ax. A similar line of reasoning gives the derivative of cos ax as $-a$ sin ax.

JFET transconductance

When a JFET is operating in the pinch-off mode, the drain current i_D is:

$$i_D = I_{DSS}\left(1 - \frac{v_{GS}}{V_P}\right)^2 \qquad (1)$$

where v_{GS} is the drain-source voltage, V_P is the pinch-off voltage and I_{DSS} is the drain current when $V_{GS} = 0$.

Transconductance is defined as:

$$g_m = \frac{di_D}{dv_{GS}}$$

In words, it is the ratio between the change of output current to change in the input voltage required to produce it. To find di_D/dv_{GS}, differentiate equation (1). This is a composite function, so use the Chain Rule:

Inner function: $\quad u = 1 - \dfrac{v_{GS}}{V_P} \quad \Rightarrow \quad \dfrac{du}{dv_{GS}} = \dfrac{-1}{V_P}$

Outer function: $\quad i_D = I_{DSS}u^2 \quad \Rightarrow \quad \dfrac{di_D}{du} = I_{DSS} \cdot 2u$

Applying the Chain Rule: $\quad \dfrac{di_D}{dv_{GS}} = \dfrac{di_D}{du} \times \dfrac{du}{dv_{GS}} = I_{DSS} \cdot 2u \cdot \dfrac{-1}{V_P}$

Substituting the value of u:

$$g_m = \frac{di_D}{dv_{GS}} = \frac{-2I_{DSS}}{V_P}\left(1 - \frac{v_{GS}}{V_P}\right) \qquad (2)$$

From equation (1), the expression in brackets equals $\sqrt{I_D}/\sqrt{I_{DSS}}$. Substituting this in equation (2):

$$g_m = \frac{-2\sqrt{I_{DSS}}}{V_P} \cdot \sqrt{I_D}$$

Given that I_{DSS} and V_P are constant, transconductance is proportional to the square root of i_D. For example, if $g_m = 2$ mA V^{-1} when $i_D = 1$ mA, then, when $i_D = 16$ mA:
$$g_m = 2 \times \sqrt{16} = 8 \text{ mA } V^{-1}$$
This result shows how transconductance varies with the instantaneous value of the drain current, causing distortion of the signal.

Test yourself 9.5

Use the Chain Rule to differentiate these functions

1 $y = (3x - 5)^2$
2 $y = (3x + 2)^5$
3 $y = 5(x + 3)^3$
4 $y = (2x^2 - 3x + 7)^2$
5 $y = \dfrac{1}{(x^3 + 2)}$
6 $y = \sqrt{3x - 2}$
7 $y = \sin 5\theta$
8 $y = \ln 6x$

Differentiation of a product

As an example, consider the function:

$$y = 3x^3(2x + 6)$$

This is the product of two functions of x:

$$3x^3 \quad \text{and} \quad 2x + 6$$

The direct way of calculating the derivative is first to multiply the two functions together, to obtain a single function:

$$y = 3x^3(2x + 6) = 6x^4 + 18x^3$$

Differentiating this gives:

$$\frac{dy}{dx} = 24x^3 + 54x^2$$

Unfortunately, this technique may lead to lengthy calculations. Also, there are functions such as $\sin x \cdot \cos x$ which cannot be multiplied out easily, and so cannot be treated in this way.

More techniques for differentiation

Function of a function (Chain Rule):

$$\frac{dy}{dx} = \frac{dy}{du} \cdot \frac{du}{dx}$$

Product of two functions:

$$\frac{dy}{dx} = \frac{du}{dx} \cdot v + \frac{dv}{dx} \cdot u$$

Quotient of two functions:

$$\frac{dy}{dx} = \frac{\frac{du}{dx} \cdot v - \frac{dv}{dx} \cdot u}{v^2}$$

A general rule for differentiation of products, called the **Product Rule**, is as follows. First write out the two functions separately, calling them u and v. In the example above:

$$u = 3x^3 \quad \text{and} \quad v = 2x + 6$$

Find their derivatives:

$$\frac{du}{dx} = 9x^2 \quad \text{and} \quad \frac{dv}{dx} = 2$$

The rule for finding the derivative of their product is given in the box. In the example:

$$\frac{dy}{dx} = 9x^2(2x + 6) + 2 \cdot 3x^3$$

$$= 18x^3 + 54x^2 + 6x^3$$

$$= 24x^3 + 54x^2$$

This gives the same result as the direct technique.

Differentiating a quotient

This rule (see box) applies when the expression to be differentiated consists of a quotient of two functions. For example:

$$y = \frac{x^2}{2x + 1}$$

Write out the two functions separately:

$$u = x^2 \quad \Rightarrow \quad \frac{du}{dx} = 2x$$

$$v = 2x + 1 \quad \Rightarrow \quad \frac{dv}{dx} = 2$$

Matching impedances

A generator, with emf E and internal resistance R_G is connected to a load resistor R_L (Figure 9.11). The power P_L transferred to the load is:

$$P_L = E^2 \cdot \frac{R_L}{(R_L + R_G)^2}$$

It is required to calculate the rate of change of power for a given change of R_L:

$$\frac{dP_L}{dR_L}$$

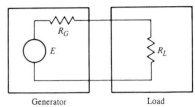

Figure 9.11

Let $u = E^2 R_L$
and $v = (R_L + R_G)^2$
$= R_L^2 + 2R_L R_G + R_G^2$

Note that v is a function, not a voltage. Differentiating u and v with respect to R_L (R_G is a constant):

$$du/dR_L = E^2$$

$$dv/dR_L = 2R_L + 2R_G = 2(R_L + R_G)$$

Applying the Quotient Rule:

$$\frac{dP}{dR_L} = \frac{(R_L + R_G)^2 \cdot E^2 - E^2 R_L \cdot 2(R_L + R_G)}{[(R_L + R_G)^2]^2}$$

Dividing throughout by $(R_L + R_G)$:

$$= \frac{E^2(R_L + R_G - 2R_L)}{(R_L + R_G)^3} = \frac{E^2(R_G - R_L)}{(R_L + R_G)^3}$$

To find when the transfer of power is a maximum, locate the point or points of inflection (p. 70) by putting the derivative equal to zero:

$$\frac{E^2(R_G - R_L)}{(R_L + R_G)^3} = 0$$

$$\Rightarrow \quad R_G - R_L = 0$$

$$\Rightarrow \quad R_L = R_G$$

There is either a maximum or a minimum transfer of power when the generator and load impedances are equal.

The derivative is too complicated to differentiate again but, by examining how its value changes in the region of $R_L = R_G$, it can be shown that this point is a *maximum*. The maximum transfer of power occurs when generator and load impedances are equal. This is an instance of the importance of impedance matching.

Apply the rule:

$$\frac{dy}{dx} = \frac{2x \cdot (2x + 1) - 2 \cdot x^2}{(2x + 1)^2}$$

$$= \frac{2x(x + 1)}{(2x + 1)^2}$$

> **Test yourself 9.6**
>
> Differentiate these products.
> 1 $(2x - 5)(3x + 4)$ 2 $(x^3 + 1)(2x - 5)$
> 3 $(2x^3 - 4x^2)(x^5 - 3)$
> 4 $x^3(3x^2 + 1)^3$ [Hint: this requires the Chain Rule too.]
>
> Differentiate these quotients.
> 5 $\dfrac{x + 1}{x - 2}$ 6 $\dfrac{4x - 7}{2x + 3}$
> 7 $\dfrac{x + 3}{x^3}$ 8 $\dfrac{2x + 1}{\sqrt{x}}$
> 9 $\dfrac{(2x + 3)^2}{5x}$ [Hint: this requires the Chain Rule too.]

Partial differentials

It may happen that a function has two variables instead of only one. If z is the function and the two variables are x and y, we show this in function notation by writing:

$$z = f(x, y)$$

For example:

$$z = f(x,y) = 2x^2 + 3xy - 4y^3$$

If we assume for the moment that y is a constant, this function can be differentiated with respect to x:

$$\frac{dz}{dy} = 4x + 3y$$

However, to indicate that this is only *part* of the way in which z may vary, we indicate that this is a **partial differential** by using the 'curly d' notation:

$$\frac{\partial z}{\partial x} = 4x + 3y$$

Similarly, we can hold x constant and partially differentiate z with respect to y:

$$\frac{\partial z}{\partial y} = 3x - 12y^2$$

This idea can be extended to three or even more variables, differentiating z with respect to any *one* of them while the others are assumed to be constant.

If $z = f(x, y)$, and each of x and y are changed by small amounts of Δx and Δy, there is a corresponding small change in z, which we call Δz. Note that the changes may be positive or negative. The amount of change in z due to the change of x depends on:

- how much x changes (the value of Δx); this is the basic cause of the change in z, and
- the rate of change of z with respect to x (the value of $\partial z/\partial x$), in other words to what extent z is affected by changes in x.

The eventual change in z is the product of these quantities, so the amount of change is:

$$\frac{\partial z}{\partial x} \cdot \Delta x$$

Similarly, the amount of change in z due to changes in y is:

$$\frac{\partial z}{\partial y} \cdot \Delta y$$

The change in z due to changes in both x and y is:

$$\Delta z = \frac{\partial z}{\partial x} \cdot \Delta x + \frac{\partial z}{\partial y} \cdot \Delta y$$

This is an approximate result, because Δx and Δ are both finite.

The rate of change, $\partial z/\partial x$, may vary over the range x to $x + \Delta x$, just as the slope of the curve dy/dx in Figure 9.3 varies along the segment from A to B. As with Figure 9.3, the result becomes more precise the smaller the value of Δx. The same applies to $\partial z/\partial y$ and the value of Δy. Given small enough values of Δx and Δy, as in the next example, the results are close enough for many purposes.

Example
Current I flows through a resistor R when a p.d. V is applied across it. To begin with, $R = 100\ \Omega$ and $V = 12$ V. By how much does I change if R falls by $2\ \Omega$ and V increases by 0.1 V?

The function relating I, R and V is:

$$I = f(R, V) = \frac{V}{R}$$

This is the familiar Ohm's Law equation in another guise. Changes in R and V are relatively small: $\Delta R = -2$, $\Delta V = 0.1$. These are small enough to allow the equation to be used. Re-writing the equation in terms of I, R and V:

$$\Delta I = \frac{\partial I}{\partial R} \cdot \Delta R + \frac{\partial I}{\partial V} \cdot \Delta V$$

Calculate the partial differentials, in each case assuming the other variable to be constant:
$$\frac{\partial I}{\partial R} = \frac{-V}{R^2} \quad \text{and} \quad \frac{\partial i}{\partial V} = \frac{1}{R}$$

Substituting:
$$\Delta I = \frac{-12}{100^2} \cdot -2 + \frac{1}{100} \cdot 0.1$$
$$= 0.0024 + 0.001 = 0.0034$$

The current increases by 3.4 mA.

Reasoning of this kind can be applied to the calculation of changes in other electronic circuits. The equation is not limited to two quantities. It is extended by adding further terms of the form:
$$\frac{\partial z}{\partial p} \cdot \Delta p$$

where p is any other electrical quantity.

Percentage changes

The equation derived above can also be used for calculations made on a percentage basis, provided that the percentages are small.

Example
A potential divider (Figure 6.2) consists of two resistors, R_A, resistance $a = 100\,\Omega$, and R_B, resistance $b = 400\,\Omega$. We assume that the divider is connected to a load of very high impedance so that current flowing to the load may be ignored. The tolerance of R_A is $\pm 1\%$ while that of R_B is $\pm 0.5\%$. By what percentage is the output likely to fall below its nominal value?

Given input V, the nominal output is:
$$v = \frac{Vb}{a + b}$$
$$= \frac{V \times 400}{500} = 0.8V$$

A fall of output results when R_B is less than its nominal value; the largest possible difference is 0.5% of b, so $\Delta b = -0.005b$. Conversely, output falls when R_A is more than its nominal value, which, with 1% tolerance, is when $\Delta a = 0.01a$.

Now to calculate the partial differential for variations in a. The partial derivative of $\frac{Vb}{a+b}$ with respect to a is:
$$\frac{\partial v}{\partial a} = \frac{-Vb}{(a+b)^2}$$

This differential is calculated using the rule for quotients (page 130). Similarly, for b:
$$\frac{\partial v}{\partial b} = \frac{Va}{(a+b)^2}$$

Note the difference in sign, reflecting the opposite effects of changes in R_A and R_B. The equation for Δv is:
$$\Delta v = V\left(\frac{\partial v}{\partial a} \cdot \Delta a + \frac{\partial v}{\partial b} \cdot \Delta b\right)$$
$$= V\left(\frac{-400 \times 0.01 \times 100}{500^2} + \frac{100 \times -0.005 \times 400}{500^2}\right)$$

Taking out factors 400 and 500:

$$\Delta v = \frac{V \times 400}{500} \left(\frac{-0.01 \times 100}{500} + \frac{100 \times -0.005}{500} \right)$$

Substitute v for 0.8 V

$$\Delta v = v \left(-\frac{1}{500} - \frac{1}{10\,000} \right)$$

$$= v \times -0.003$$

The fall in v is 0.3%.

The equation for the divider is a simple one so we can easily check the result by substituting the extreme values of a and b directly into the equation. If $V = 100$ V, then nominally $v = 80$ V. With $a = 101\,\Omega$ and $b = 398\,\Omega$:

$$v = 100 \times \frac{398}{499} = 79.760 \text{ V (3 dp)}$$

This is a fall of 0.24 V, which is 0.3% of 80 V, as required. The result is so easy to check that one might wonder why it is necessary to bother with partial differentials. One reason for using them is that they provide an equation which can be used with any combination of voltages and tolerances. Secondly, the above is just an elementary example to illustrate the method. In the majority of circuits, the relationships are more complicated and partial differentiation is the simplest technique.

Rates of change

It is common in circuits for quantities to vary with time. In other words they have a *rate of change*. If two such changing quantities both affect a third quantity, we use partial differentials to define their relationship. In terms of partial differentials, a quantity z is a function of two quantities x and y, as before, but now both x and y are functions of time.

We start with the equation for the sum of the changes in z, as derived above:

$$\Delta z = \frac{\partial z}{\partial x} \cdot \Delta x + \frac{\partial z}{\partial y} \cdot \Delta y$$

This is based on changes taking place during an unspecified length of time. If we divide all terms in the equation by Δt, we obtain the changes occurring during a definite but short length of time Δt:

$$\frac{\Delta z}{\Delta t} = \frac{\partial z}{\partial x} \cdot \frac{\Delta x}{\Delta t} + \frac{\partial z}{\partial y} \cdot \frac{\Delta y}{\Delta t}$$

In the limit, as Δt approaches zero:

$$\frac{dz}{dt} = \frac{\partial z}{\partial x} \cdot \frac{dx}{dt} + \frac{\partial z}{\partial y} \cdot \frac{dy}{dt}$$

This is the **total differential** of z with respect to t. Given the functions which tell us how x and y vary with t, and how z depends on x and y, we can calculate how z varies with t. Note that it does not contain terms such as Δx or Δy, and therefore gives a precise result.

Example
A ramp generator circuit in which the output voltage v is increasing at the rate of $15\,\text{V}\,\text{s}^{-1}$ supplies current to a resistor which is decreasing at a rate of $2\,\Omega\,\text{s}^{-1}$. The instantaneous current i is given by $i = v/r$. What is the rate of change of current when $v = 5\,\text{V}$ and $r = 450\,\Omega$?

Find the partial differentials. From the example on page 133:

$$\frac{\partial i}{\partial r} = \frac{-v}{r^2} \quad \text{and} \quad \frac{\partial i}{\partial v} = \frac{1}{r}$$

Substitute these and the values of the rates of change in the total differential equation:

$$\frac{di}{dt} = \frac{-5}{450^2} \cdot -2 + \frac{1}{450} \cdot 15$$

$$= 0.000\,049\,38 + 0.033\,33$$

$$= 0.033\,4 \; (3\,\text{sf})$$

The current is increasing at the rate of $33.4\,\text{mA}\,\text{s}^{-1}$.

Another factor which produces changes in circuits is temperature. The characteristics of many types of component, including resistors, capacitors and semiconductors have a temperature coefficient (tempco), usually rated in parts per million per degree Celsius. Thus voltages, currents, resistances and capacitances may all be functions of temperature. The equation for the effect of temperature on circuit performance is the same as that given above, except that the symbol t refers to temperature instead of time. As mentioned on page 134, the equation can be extended to include as many quantities as required, simply by adding extra terms, in this case in the form:

$$\frac{\partial z}{\partial p} \cdot \frac{dp}{dt}$$

where p is the quantity concerned.

Test yourself 9.7

Find the partial differentials of the following functions with respect to x and to y.

1. $z = 5x^2 + 2xy - y^2$
2. $z = (3x^2 - y)(2x + 5y^2)$
3. $z = \sin(2x - 3y)$
4. A 47 Ω resistor has a p.d. of 100 V across it. The power P being dissipated in the resistor is given by:

$$P = \frac{V^2}{R}$$

If the voltage increases by 2 V and the resistance falls by 0.5 Ω, by what amount does the power change? (to 3 sf)

5. In a voltage-regulating circuit (Figure 9.12) a constant voltage $V = 12$ is applied. The band-gap reference produces a reduced voltage v with a tempco of 70 ppm per °C. The resistor has a tempco of 100 ppm per °C. At a temperature of 25°C, $v = 5V$ precisely and $r = 150\,\Omega$. The current through the resistor is

$$i = \frac{V - v}{r}$$

What is the tempco of the current at 25°C?

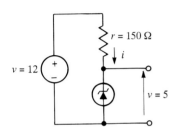

Figure 9.12

10 Summing it up

> From totalling the charge accumulating on a capacitor to analysing the waveform of signals, the topic of this chapter is fundamental to many aspects of electronics.
> You need to know about differentiating (Chapter 9), logarithms (page 25) and exponentials (page 59).

An interesting relationship exists between the curves of a function and of its derivative. Take this function as an example:

$$y = x^2 + 10$$

This is the upper curve drawn in Figure 10.1. Below it is the curve for its derivative:

$$\frac{dy}{dx} = 2x$$

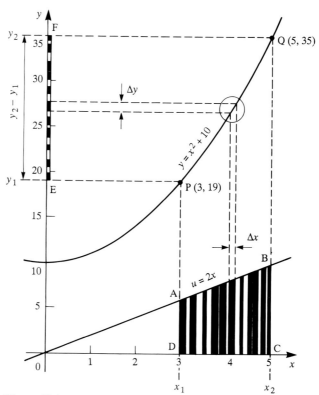

Figure 10.1

The derivative is also a function and, to simplify the discussion, we call this function u:

$$\frac{dy}{dx} = u$$

There is a region ABCD beneath the graph of u, from $x_1 = 3$ to $x_2 = 5$. The aim is to calculate the area A of this region.

If the graph is a straight line, as it is here, the area is easily calculated by using the formula for a trapezium:

$$A = CD \times \frac{(AC + BD)}{2} = 2 \times \frac{(6 + 10)}{2} = 16$$

This gives us the result we are aiming for, but we need a more general method for finding A, which works with a graph of any function.

Think of the area as being divided into a large number of narrow vertical strips. In the figure these are alternately black and white to make them show up clearly. Given that the width of a strip is Δx (a very small distance in the x- direction) and that its height is u, the area of a strip is approximately:

Area of strip = $u\Delta x$.

The total area of ABCD is the sum of the areas of all the strips. In maths, this is written:

$$A = \sum_{x_1}^{x_2} u\Delta x \qquad (1)$$

The symbol 'Σ' means 'sum all terms similar to the following term'. The 'x_1' and 'x_2' indicate that this is to be done for the area starting at x_1 and finishing at x_2.

Now we leave the strips for the moment and look at the line EF at the top left of the figure. This is divided into line segments, alternately black and white. The segments in order from E to F correspond with the strips in order from AD to BC. The figure shows the correspondence for one of the strips. Its width is projected up on to the function curve, and then across on to EF to give the length of the segment. The length of a segment is Δy, a very small distance in the y-direction. Since they correspond with the segments of varying widths and since the curve is not straight, the segments are not necessarily of equal length.

The total length of line EF is the sum of the lengths of all the segments. This is written:

$$EF = \sum_{y_1}^{y_2} \Delta y \qquad (2)$$

At this stage there are two equations, (1) for the total area ABCD and (2) for the total length EF. The final step is to link these together. Figure 10.2 shows how this is done. This is an enlargement of the part of the curve circled in Figure 10.1.

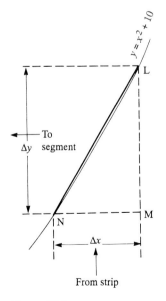

Figure 10.2

The gradient of the curve is approximately that of the straight line NL:

$$\text{Gradient} = \frac{LM}{NM} = \frac{\Delta y}{\Delta x}$$

But the gradient of the curve is also given by the value of the derivative, u:

$$\Rightarrow \quad \frac{\Delta y}{\Delta x} = u$$

$$\Rightarrow \quad \Delta y = u\Delta x$$

Summing both sides of this equation for all segments and strips:

$$\sum_{y_1}^{y_2} \Delta y = \sum_{x_1}^{x_2} u\Delta x$$

Looking back at the summing equations (1) and (2), we see that this equation means:

The length EF = the area of ABCD

This is an extremely important result. To show that it works in this case, for an area from $x_1 = 3$ to $x_2 = 5$, we use the function $y = x^2 + 10$ to calculate the corresponding values of:

$$y_1 = 19 \quad \text{and} \quad y_2 = 35$$

The length EF = $y_2 - y_1$ = 35 − 19 = 16. This is equal to the result calculated from the trapezium formula.

Tidying up

Two approximations were made during the discussion above:

- the area of the strips was calculated as if they are rectangular, but they are not;
- the gradient of the curve (Figure 10.2) was calculated as if it is the same as the straight line LN, which it is not.

These approximations gradually disappear as Δx and Δy are made smaller and smaller. As the strips become narrower they become less and less different from rectangles. The line LN becomes less and less different from the curve. Taking the calculations to the limit (as Δx and Δy approach zero) removes the errors due to the approximations. When working at the limit, with Δx and Δy infinitely small, the Greek letter S, or *sigma* Σ, which stands for 'sum', is replaced by a 'long S'. Equation (2) becomes:

$$A = \int_{y_1}^{y_2} dy$$

The symbol Δy is replaced by 'dy', just as in differentiation (page 114), to indicate that limiting values are involved.

Another word meaning almost the same as a summation is **integration**, 'to make into a whole'. This is done to individual components when an integrated circuit (IC) is made. We have done it here, putting the strips together to make a whole area, or putting the segments together to make a whole line. The expression above is referred to as an **integral**. It is the 'integral with respect to y between the limits y_1 and y_2'. In this sentence the term 'limit' is being used in its other sense (page 103); it means 'the lower and upper values of y'.

Working in reverse

The result of the discussion of integration can be expressed as follows:

Given the curves of the two functions, one of which is the derivative of the other, the area under the curve of the derivative from x_1 to x_2 is given by the difference of the y-coordinates ($y_2 - y_1$) of the other curve.

In practical problems, this line of argument is usually looked at from an opposite point of view. We are given a function and want to know the area under a segment of its curve. The function we are *given* is the *derivative*. Before we can calculate the values of y_2 and y_1 we have to find a function which has this derivative. In other words, we have to differentiate in reverse. Integration is sometimes called **anti-differentiation**, though the name *integration* is a reminder that the process is concerned with summing small parts to make a whole.

On the other hand, the name *anti-differentiation* implies that the process is one based on the reverse of the differentiation rules. The box shows how to integrate a simple expression of the form ax^n.

Rules for integrating simple functions

Provided that the index of x is not -1:

Add 1 to the index
Divide the term by the *new* index
Add c, the constant of integration

Summarizing:

$$y = ax^n \quad \Rightarrow \quad \int y \, dx = \frac{ax^{n+1}}{n+1} + c$$

$$n \neq -1$$

If the function to be integrated has a constant as multiplier or divisor, that constant applies to every one of the strips that are being integrated. It therefore applies to the *whole* area and so may be written in front of the integral sign. For example

$$\int 5x^2 \, dx = 5 \int x^2 \, dx$$

$$\int \frac{x^3}{2} \, dx = \frac{1}{2} \int x^3 \, dx$$

Comparing these rules with those in the box on page 118 shows the inverse nature of integration.

The rules for integration are simply the inverse of the rules for differentiation, with the extra condition that we have to add an unknown constant, c, the **constant of integration**. Given a polynomial with a constant term *not* involving x, we lose that term when differentiating. For example:

$$y = 2x^2 + 3x$$
$$y = 2x^2 + 3x + 4$$
$$y = 2x^2 + 3x + 7 \quad \text{and}$$
$$y = 2x^2 + 3x + 99$$

all differentiate to

$$dy/dx = 4x + 3$$

The constant makes no difference to the value of the derivative. When reverse differentiating $4x + 3$, there is no way of knowing whether the constant in the original function was 4, 7, 99 or any other value, or even if there was a constant at all. We just call it c, with the possibility that $c = 0$. More about this later.

Examples of using the rules for integration:

Function	Integral
$2x$	$x^2 + c$
$4x^2$	$\dfrac{4x^3}{3} + c$
$6x^2 + 7x$	$2x^3 + \dfrac{7x^2}{2} + c$
$5/x^2 \ (= 5x^{-2})$	$-5/x \ (= -5x^{-1}) + c$
$2/x^4 \ (= 2x^{-4})$	$-2/3x^3 \ \left(= \dfrac{-2x^{-3}}{3}\right) + c$
$3\sqrt{x} \ (= 3x^{1/2})$	$2(\sqrt{x})^3 \ (= 2x^{3/2}) + c) + c$

These examples can be checked by differentiating the functions in the right column to yield the functions in the left column. Both fractional and negative indices follow the rules, the only exception being the index -1, in a term such as $x^{-1} \ (= 1/x)$. Applying the rules, the new index is 0, and the term has to be divided by 0, giving an indeterminate result. A glance at the table on page 118 supplies the answer. The function which has $1/x$ as its derivative is $\ln x$. Therefore the integral if $1/x$ is $\ln x + c$.

Standard integrals

The box lists some of the more commonly occurring functions and their integrals. When given a function to integrate, the first step is to examine it to see if it is in the form of one of the standard integrals. The examples given in the previous paragraph are all variations of the standard forms. The terms of a polynomial are integrated individually, as in one of the examples above.

Standard integrals

(add c to all integrals)

Function	Integral	Example	
		Function	Integral
ax^n	$\dfrac{ax^{n+1}}{n+1}$	$3x^4$	$\dfrac{3x^5}{5}$
a/x	$a \ln x$	$5/x$	$5 \ln x$
$\sin ax$	$\dfrac{-\cos ax}{a}$	$\sin 5x$	$\dfrac{-\cos 5x}{5}$
$\cos ax$	$\dfrac{\sin ax}{a}$	$\cos 3x$	$\dfrac{\sin 3x}{3}$
e^{ax}	$\dfrac{e^{ax}}{a}$	e^{2x}	$\dfrac{e^{2x}}{2}$

If an expression (or **integrand**, as we call a function that is to be integrated) is not in standard form, it may be possible to convert it to standard form before integrating it. For example, given:

$$\int x(2x - 6)\, dx$$

The integrand is the *product* of two functions. If the expression is multiplied out, we obtain a polynomial, the terms of which can be integrated individually as standard integrals:

$$\int x(2x - 6)\, dx = \int (2x^2 - 6x)\, dx$$
$$= \frac{2x^3}{3} - 3x^2 + c$$

The integral of a quotient may sometimes be found by first dividing out the integrand:

$$\int \frac{x^3 + 1}{x^2}\, dx = \int \left(x + \frac{1}{x^2}\right) dx$$
$$= \frac{x^2}{2} - \frac{1}{x} + c$$

Not all integrands lend themselves to this approach. Further methods of integration are described later in this chapter.

Dealing with the constant of integration

In the examples above, the integral has had no limits attached to it. Such an integral is called an **indefinite integral**. It is evaluated by substituting a given value of x (or other variable that the integral is in respect to). But there is still the constant of integration to be accounted for. To find this, further information is required, such as the value of the integral for a given value of x.

Example
Evaluate the integral of $4x + 3$, given that, when $x = 2$ the value of the integral is 10.

$$\int (4x + 3)\,dx = 2x^2 + 3x + c$$

When $x = 2$:

$$\int (4x + 3)\,dx = 2 \cdot 2^2 + 3 \cdot 2 + c = 14 + c$$

Put this equal to the given value

$$14 + c = 10$$
$$\Rightarrow \quad c = -4$$

The integral is $2x^2 + 3x - 4$

In the earlier discussion of integration, the area under the curve (Figure 10.1) was to be evaluated between the limits, x_1 and x_2. Such an integral with limits is called a **definite integral**:

$$A = \int_{x_1}^{x_2} 2x\,dx = [x^2 + c]_{x_1}^{x_2}$$

The integral is written in square brackets, with the limits just outside the right bracket. Evaluating the contents of the brackets for $x = x_1$ and $x = x_2$ gives y_1 and y_2; their difference, $y_2 - y_1$, is the area (page 140):

$$A = [x_2^2 + c] - [x_1^2 + c]$$
$$= x_2^2 - x_1^2$$

Given $x_1 = 3$ and $x_2 = 5$, as on page 139:

$$A = 5^2 - 3^2 = 25 - 9 = 16$$

Note that c appears in *both* brackets and so cancels out. Consequently, when we are evaluating definite integrals, the constant of integration can be ignored. In Figure 10.1 it can be seen that the value of c is 10, this being the y-intercept of the curve $y = x^2 + 10$. However, it does not matter what it is. Giving c a different value merely shifts the parabola vertically up or down the page, without affecting the length of EF.

When integrating the function $1/x$, the constant of integration can be expressed in a different way. Normally we would write:

$$\int \frac{1}{x} = \ln x + c$$

Using an alternative expression, we write:

$$\int \frac{1}{x} = \ln kx$$

The constant, now called k, is included in the logarithm. The basis of this is that addition of logarithms is equivalent to the multiplication of ordinary numbers:

$$\ln kx = \ln k + \ln x$$

differentiating:

$$\frac{d}{dx}(\ln kx) = \frac{d}{dx}(\ln k + \ln x) = 0 + \frac{1}{x} = \frac{1}{x}$$

Conversely the integral of $1/x$ can be considered to be $\ln kx$. This form is useful in differential equations, as will be seen in a later chapter.

Integrating with time

A constant current I flows into a capacitor, capacitance C, for a period of time t. Assuming that the capacitor has no charge at the instant timing begins, the charge q stored in the capacitor is given by:

$$q = It$$

In words, the charge is the *product* of the current and the length of time for which it has been flowing. We can represent this as a graph (Figure 10.3a) in which charge, being the product of current and time, is represented by the *area* beneath the curve. This idea can be extended to a varying current i since, for any short instant of time Δt, the accumulating charge Δq equals $i\Delta t$ (Figure 10.3b); compare with the strips of Figure 10.1. Now suppose that we pass a current which varies according to a given function in which t is the independent variable. The area under the current–time graph represents the accumulated charge. This area is found by evaluating the definite integral of the function for the period of time concerned.

Example
A capacitor begins uncharged, but is then charged by a current i for which $i = \sin 20t$. How much charge accumulates on the capacitor in a period of 1.2 s? Integrate the function i with respect to t, from $t_1 = 0$ to $t_2 = 1.2$ (4 dp):

$$q = \int_{t_1}^{t_2} \sin 20t \, dt = \left[\frac{-\cos 20t}{20}\right]_{t_1}^{t_2}$$

$$= \left[\frac{-\cos(20 \cdot 1.2)}{20}\right] - \left[\frac{-\cos 0}{20}\right]$$

$$= -0.0212 - (-0.0500)$$

$$= 0.0288$$

The charge is 0.0288 C.

Note that the angle $20t$ after 1.2 s is 24 rad. This is $24/2\pi = 3.82$ cycles. The capacitor is subject to charging and discharging three times, finishing with 0.0288 C, when 0.82 of the way through the fourth cycle.

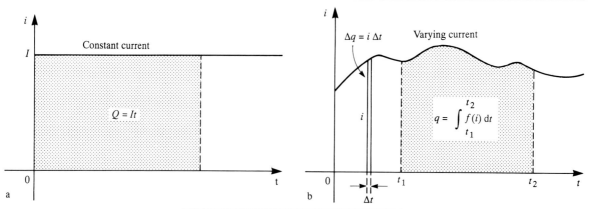

Figure 10.3

Test yourself 10.1

Find these indefinite integrals.

1 $\int (x^2 - 2x)\, dx$
2 $\int (x^3 - 3x + 2)\, dx$
3 $\int 7\, dx$
4 $\int (x + 2)(x + 3)\, dx$
5 $\int 3x(1 + 2x)\, dx$
6 $\int \dfrac{1}{\sqrt{x}}$
7 $\int \cos 3x\, dx$
8 $\int (\sin\theta + 3)\, d\theta$
9 $\int -2 \cos 3\theta\, d\theta$
10 $\int e^{3x}\, dx$
11 $\int e^x (e^{3x} - 2e^x)\, dx$
12 $\int \dfrac{7}{4x}\, dx$

Find the exact integral of the following.

13 $\int (2x + 4)\, dx$, given that the value of the integral is 6 when $x = 1$.

14 $\int (x + 3)(3x - 5)\, dx$, given that the value of the integral is -11 when $x = 2$.

15 $\int \dfrac{x^2 - 1}{x^4}\, dx$, given that the value of the integral is 8 when $x = 1/3$.

[Hint: split it into two fractions.]

Find these definite integrals (4 sf).

16 $\int_1^4 (x^2 + 2)\, dx$
17 $\int_4^{16} \sqrt{x}\, dx$
18 $\int_2^4 (2x^3 - 3x^2)\, dx$
19 $\int_0^\pi \sin x\, dx$
20 $\int_1^3 (2x - 3)^2\, dx$
21 $\int_0^1 e^{2x}\, dx$

Integrating ratios

If the integrand is in the form of a quotient, it may be possible to integrate it after dividing out the quotient. For example:

Find $\displaystyle\int \frac{2x + 3}{x + 2}\, dx$

Divide $(2x + 3)$ by $(x + 2)$:

$$\begin{array}{r}
2 \\
x + 2 \overline{\smash{)}\, 2x + 3} \\
2x + 4 \\ \hline
-1
\end{array}$$

$$\frac{2x + 3}{x + 2} = 2 - \frac{1}{x + 2}$$

$$\int \left(2 - \frac{1}{x + 2}\right) dx = 2x - \ln(x + 2) + c$$

Note that the logarithm of a negative number is indeterminate. If the expression $(x + 2)$ has a negative value, the logarithm of this can not be found. For this reason, it is the *absolute value* of the expression which must be used when evaluating the logarithm. The result of the integration should be written:

$$2x - \ln|x + 2| + c$$

Integration by substitution

This is a technique to be tried when an integrand is not one of the standard integrals, and cannot be simplified or multiplied out to make it into a standard integral. The idea of this technique is to put everything in terms of a different variable, with the aim of making it easier to integrate. Take an example:

Find $\int (4x - 3)^3\, dx$

This *could* be integrated by expanding it into a polynomial and then integrating the terms individually, but we will use the substitution method instead. The variable of the new integral is to be u, and the first step is to replace the integrand by u, by making:

$$u = 4x - 3 \qquad (1)$$

The choice of what to make u is straightforward in this example, though it is not always as easy. Usually it is best to try substituting for the most complicated part of the integrand and, if this fails to give a solution, to try other parts. The original integral also has 'dx' in it, so we must find a substitute for this, in terms of u. Differentiating equation (1) gives:

$$\frac{du}{dx} = 4$$

Although we said on page 114 that dy/dx (or dx/du, in this case) is a symbol expressing limiting rates of change, implying that it must always be written as a *whole*, we find in practice that we can separate the dx from the du, treating them as *individual quantities*. This is why we can change the subject of equation (2) to provide the required substitute for dx:

$$dx = \frac{du}{4}$$

Now we are ready to assemble the new integral by substituting the equivalents of $(4x - 3)^4$ and dx:

$$\int (4x - 3)^3 \, dx = \int u^3 \, \frac{du}{4}$$

$$= \frac{1}{4} \int u^3 \, du$$

$$= \frac{u^4}{4 \times 4} + c = \frac{u^4}{16} + c$$

Integrating by substitution

Put u equal to (one of) the function(s) of x.

Find du/dx or dx/du (whichever is easier), then change subject to get substitute for du.
(If necessary, find a substitute for other term in x.)

Assemble the new integral by substitution.
(Simplify, if possible.)

Integrate.

Replace u's with original function of x.
(Simplify, if possible.)

This gives the integral in terms of u but we need it in terms of the original variable x. Use equation (1) to replace u by $(4x - 3)$:

$$\int (4x - 3)^3 \, dx = \frac{(4x - 3)^4 + c}{16}$$

The integral above has the form of $ax + b$, with $a = 4$ and $b = -3$. The box shows a general rule which the result above confirms. This rule applies to any of the standard forms. Examples are:

$$\int \sin(3x + 2) \, dx = -\frac{1}{3} \cos(3x + 2) + c$$

$$\int e^{5x - 2} \, dx = \frac{1}{5} e^{5x - 2} + c$$

$$\int \frac{1}{6x + 2} \, dx = \frac{1}{6} \ln|6x + 2| + c$$

Putting ($ax + b$) into the standard forms

Instead of x, write $ax + b$ in the standard integral.
Multiply the result by $1/a$.

More examples of substitution

The next example is slightly different from the previous one:

Find $\int x(2x+3)^4 \, dx$ (1)

We have *two* functions in the integrand, x and $(2x+3)^4$. As stated above, the best approach is to try putting u equal to the more complicated function:

$u = 2x + 3$ (2)

Find a substitute for dx:

$$\frac{du}{dx} = 2 \quad \Rightarrow \quad dx = \frac{du}{2}$$

In this example we also need a substitute for the first function of the integrand, the x. This can be obtained by changing the subject of equation (2):

$$x = \frac{u-3}{2}$$

Now assemble the new integral by substituting the equivalents of x, $(2x+3)^4$ and dx:

$$\int x(2x+3)^4 \, dx = \int \left(\frac{u-3}{2}\right) u^4 \frac{du}{2}$$

$$= \frac{1}{4} \int (u-3) u^4 \, du$$

Multiplying out:

$$= \frac{1}{4} \int (u^5 - 3u^4) \, du$$

$$= \frac{1}{4} \left(\frac{u^6}{6} - \frac{3u^5}{5}\right) + c$$

$$= \frac{u^6}{24} - \frac{3u^5}{20} + c$$

Replacing terms in u with their equivalents in x:

$$= \frac{(2x+3)^6}{24} - \frac{3(2x+3)^5}{20} + c$$

From now on it is just a matter of simplifying the expression, beginning by taking out the common factor $(2x+3)^5$:

$$= (2x+3)^5 \left(\frac{2x+3}{24} - \frac{3}{20}\right) + c$$

Now add fractions, over the HCF of their denominators, 120:

$$= (2x+3)^5 \left(\frac{5(2x+3) - 18}{120}\right) + c$$

$$= \frac{(2x+3)^5 (10x - 3)}{120} + c$$

Integrating by parts

This is another technique to be tried when the integrand does not conform to one of the standard patterns. The technique is based on the rule for differentiating a product, as on page 130. Under the Product Rule, if the function to be differentiated is the product of u and v, both of which are functions of x, then:

$$\frac{d}{dx}(uv) = \frac{du}{dx} \cdot v + \frac{dv}{dx} \cdot u$$

Reversing the order of the factors on the right makes no difference to the values of the terms:

$$\frac{d}{dx}(uv) = v \cdot \frac{du}{dx} + u \cdot \frac{dv}{dx}$$

Changing the subject:

$$u \cdot \frac{dv}{dx} = \frac{d}{dx}(uv) - v \cdot \frac{du}{dx} \, dx$$

Integrating both sides:

$$\int u \cdot \frac{dv}{dx} \, dx = \int \frac{d(uv)}{dx} \, dx - \int v \cdot \frac{du}{dx} \, dx$$

But

$$\int \frac{d(uv)}{dx} \, dx = uv + c$$

So

$$\int u \cdot \frac{dv}{dx} \, dx = uv - \int v \cdot \frac{du}{dx} \, dx$$

Looking through this again, we might ask what has happened to c, the constant of integration obtained when we integrate uv? This is simply amalgamated with the c which we will obtain when integrating v in the final equation.

To use this equation, we have to consider the integrand as being made up of two parts, one corresponding to u and other to dv/dx. For example:

Find $\int x \cos x \, dx$

The integrand has two parts, x and $\cos x$. We will make x equal to u, and $\cos x$ equal to dv/dx. As well as u and dv/dx, the equation above also contains du/dx and v, which we must now calculate:

If $u = x$, then $\dfrac{du}{dx} = 1$

If $\dfrac{dv}{dx} = \cos x,$

then $v = \sin x$ (integrating or antidifferentiating with respect to x).

Now substitute these values in the equation:

$$\int x \cos x \, dx = \int u \frac{dv}{dx} dx$$

$$= uv - \int v \frac{du}{dx} dx$$

$$= x \sin x - \int \sin x \, dx$$

$$= x \sin x - \cos x + c$$

Another example:

Find $\int x^2 \ln x \, dx$

If $u = x^2$, then $\frac{du}{dx} = 2x$

If $\frac{dv}{dx} = \ln x$,

there are problems, this is not a standard form which can be integrated easily.

Try the alternative approach:

If $u = \ln x$, then $\frac{du}{dx} = \frac{1}{x}$.

If $\frac{dv}{dx} = x^2$ then $v = \frac{x^3}{3}$ (a standard integral)

Substituting:

$$\int x^2 \ln x \, dx = \int u \frac{dv}{dx} dx$$

$$= uv - \int v \frac{du}{dx}$$

$$= \frac{x^3}{3} \cdot \ln x - \int \frac{x^3}{3} \cdot \frac{1}{x} dx$$

$$= \frac{x^3}{3} \cdot \ln x - \int \frac{x^2}{3} dx$$

$$= \frac{x^3}{3} \cdot \ln x - \frac{x^3}{9}$$

A third example:

Find $\frac{1}{L} \int mt e^{Rt/L} \, dt$

Where L, m and R are constants. First bring the constant m before the integral sign, then integrate by parts with:

$$u = t, \quad \text{amd} \quad \frac{du}{dt} = 1$$

$$\frac{dv}{dt} = e^{Rt/L} \quad \text{and} \quad v = \frac{L}{R} e^{Rt/L}$$

Substituting:

$$\frac{m}{L} \int te^{Rt/L} \, dt = \frac{m}{L} \left(t \cdot \frac{L}{R} e^{Rt/L} - \int \frac{L}{R} e^{Rt/L} \, dt \right)$$

$$= \frac{m}{L} \left(t \cdot \frac{L}{R} e^{Rt/L} - \frac{L}{R} \cdot \frac{L}{R} e^{Rt/L} \right)$$

$$= \frac{m}{L} \left(t \cdot \frac{L}{R} e^{Rt/L} - \frac{L^2}{R^2} e^{Rt/L} \right)$$

$$= \frac{m}{R^2} e^{Rt/L} (Rt - L)$$

The result of this calculation is used on page 167.

Test yourself 10.2

Integrate these quotients.

1. $\dfrac{4x - 5}{2x + 1}$ 2. $\dfrac{3x^2 - 1}{x - 3}$ 3. $\dfrac{2x^2 + 6x - 8}{x + 3}$

Integrate these expressions by substitution.

4. $\int (2x + 1)^5 \, dx$ 5. $\int x(3x + 1) \, dx$ 6. $\int \dfrac{x}{x + 1} \, dx$

Integrate these expressions by parts.

7. $\int x \sin x \, dx$ 8. $\int x \ln x \, dx$ 9. $\int xe^x \, dx$

Integration and averages

Quantities such as currents and voltages often vary with time. Figure 10.4a illustrates a voltage v which falls steadily with time. At time t_1 the voltage is v_1 and at time t_2 it is v_2. Since voltage is falling at a regular rate we can say that the average voltage is:

$$v_{av} = \frac{v_1 + v_2}{2}$$

One point to note is that the area under the graph of Figure 10.4a from t_1 to t_2 is the same as the area of the rectangle ABCD of Figure 10.4b; the shaded

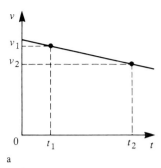

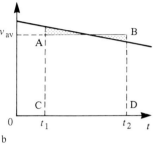

Figure 10.4

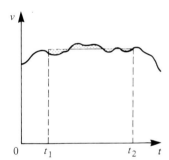

Figure 10.5

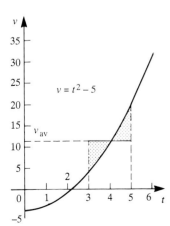

Figure 10.6

areas above and below AB being exactly equal. The area of the rectangle is:

area = height × base

In the case of rectangle ABCD:

area = $v_{av} \times (t_2 - t_1)$

Changing the subject of this equation we have:

$$v_{av} = \frac{\text{area}}{t_2 - t_1}$$

To find the average value of a varying voltage (or other variable) we only have to find the area under the graph and divide it by the length of the base. In Figure 10.4, finding the area is purely a matter of geometry but, if the line of the graph is a more complicated one, we can use integration.

Figure 10.5 shows a curve which, judged from its shape, must represent a rather complicated function. Without having to say precisely what the function is, we can say that v is a function of time, or $v = f(t)$. Once again it is possible to draw a rectangle having the same area as the area under the graph between t_1 and t_2. Once again:

$$v_{av} = \frac{\text{area}}{t_2 - t_1}$$

It is easy to calculate $t_2 - t_1$, so the main problem in calculating v_{av} is to find the area. The area is the definite integral of function $v = f(t)$ from t_1 to t_2:

$$\text{area} = \int_{t_1}^{t_2} v \, dt$$

Putting this into the equation above we arrive at an equation for v_{av}:

$$v_{av} = \frac{1}{t_2 - t_1} \int_{t_1}^{t_2} v \, dt$$

Example
Given the function $v = t^2 - 5$, calculate the mean value of v over the range $t_1 = 3$ to $t_2 = 5$. Figure 10.6 shows the curve and the area beneath it. From the equation above:

$$v_{av} = \frac{1}{5 - 3} \int_3^5 (t^2 - 5) \, dt$$

$$= \frac{1}{2} \left[\frac{t^3}{3} - 5t \right]_3^5$$

$$= \frac{1}{2} \left[\left[\frac{125}{3} - 25 \right] - \left[\frac{27}{3} - 15 \right] \right]$$

$$= \frac{1}{2} \cdot \frac{50 + 18}{3} = \frac{68}{6} = 11.33 \text{ (4 sf)}$$

The mean can also be taken over a part of the curve for which v is negative. For example, integrating from t_1 to $= 0$ to $t_2 = 5$ gives:

$$v_{av} = \frac{1}{5} \cdot \frac{50}{3} = 3.33 \text{ (3 sf)}$$

Although the range of t is greater than before, v_{av} is less than before because it now includes values of v which are negative. In terms of area, the area below the x-axis is *negative area*.

Integrating sine waves

The equation for the average value of v can be applied when v is a trig function. Taking the simplest possible case:

$$v = \sin t$$

This is the equation for a sine wave in which amplitude $A = 1$ and angular velocity $\omega = 1$ (compare the equation on page 67). The period of 1 cycle is $P = 2\pi/\omega = 2\pi$. This corresponds to a frequency of $1/2\pi = 0.16\,\text{Hz}$. We will find the average value of v during one cycle.

Applying the equation:

$$v_{av} = \frac{1}{2\pi} \int_0^{2\pi} \sin t\, dt$$

$$= \frac{1}{2\pi} [-\cos t]_0^{2\pi} = \frac{1}{2\pi} [-\cos 2\pi + \cos 0]$$

$$= \frac{1}{2\pi} [-1 + 1] = \frac{1}{2\pi} [0] = 0$$

The average value is zero because the positive values of v from 0 to π (Figure 9.9) are exactly cancelled out by the negative values of v from π to 2π. This is the same situation as we mentioned in connection with the curve of Figure 10.6.

For comparison, calculate the value of v for a half-cycle, from $t_1 = 0$ to $t_2 = \pi$. Incidentally, all the examples we have looked at relate to varying voltage but they could equally well deal with varying current and its average i_{av}. We will keep to v_{av}, to avoid confusing 'i' with '1'. For a sine-wave voltage during half a cycle:

$$v_{av} = \frac{1}{\pi} [-\cos t]_0^{\pi} = \frac{1}{\pi} [-\cos \pi + \cos 0]$$

$$= \frac{1}{\pi} [1 + 1] = \frac{2}{\pi}$$

This result applies when amplitude $A = 1$. If we include A in the equation, it becomes $v = A \sin t$. Integrating this from 0 to π gives:

$$v_{av} = 2A/\pi$$

We can adapt the equation further to cover other frequencies, when $\omega \neq 1$. The function then becomes

$$v = A \sin \omega t$$

Integrating:

$$v_{av} = \frac{1}{t_2 - t_1} \int_{t_1}^{t_2} A \sin \omega t \, dt$$

$$\Rightarrow \quad v_{av} = \frac{A}{(t_2 - t_1) \omega} \cdot [-\cos \omega t]_{t_1}^{t_2}$$

See page 148 for rules for integrating expressions of the form $(ax + b)$.

Example
Calculate the average voltage when $v = 24 \sin 50\pi t$, for the period $t_1 = 0$ to $t_2 = 0.01$ s. Examining the function, we find that $A = 24$ and $\omega = 50\pi$. Applying the equation for v_{av}:

$$v_{av} = \frac{24}{0.01 \times 50\pi} \{[-\cos (50\pi \times 0.01)] - [-\cos 0]\}$$

$$= 15.28 \, ([0] - [-1])$$

$$= 15.28$$

Modified sine waves

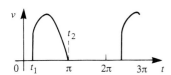

Figure 10.7

The curve shown in Figure 10.7 is one that might be the voltage output of a thyristor-controlled power supply. To simplify the calculation, let $A = 1$ and $\omega = 1$. Now we have to integrate from $t_1 = \theta$ to $t_2 = \pi$:

$$v_{av} = \frac{1}{\pi} [-\cos t]_{\theta}^{\pi} = \frac{1}{\pi} [-\cos \pi + \cos \theta]$$

$$= \frac{1}{\pi} (1 + \cos \theta)$$

This result shows that, if the voltage is switched on when $\theta = 0$, the average value is 2π, as found in the previous calculation. If θ is increased gradually from 0 to π, the average value gradually falls to zero.

Root mean square (rms) values

The **root mean square** of an alternating voltage (or current) is the square root of the mean of the squares of the instantaneous values of the voltage (or current). Because it is equal to the steady dc voltage or current which dissipates the same power in a resistance, it is an important quantity in electronics. The definition sounds involved but the calculation is similar to the calculation of the average value given above, except that we *square* the function before we integrate it, then take the *square root* of the result. In symbols:

$$v_{rms} = \sqrt{\frac{1}{t_2 - t_1} \int_{t_1}^{t_2} v^2 \, dt}$$

Compare this with the equation for v_{av} on page 153. In this expression, v is any function of t, but most practical calculations are concerned with sine waves. For a sine wave of amplitude A, angular frequency ω:

$$v = A \sin \omega t$$

To obtain an rms value we have to integrate v^2, but:

$$v^2 = A^2 \cdot \sin^2 \omega t$$

This means that we have to integrate $\sin^2 \omega t$, using the trig identity $2\sin^2\omega t \equiv 1 - \cos 2\omega t$. Without going into details, the integration yields this result:

$$v_{rms} = \sqrt{\frac{A^2}{2(t_2 - t_1)}\left[t - \frac{\sin 2\omega t}{2\omega}\right]_{t_1}^{t_2}} \qquad (1)$$

Over a whole cycle, from $t_1 = 0$ to $t_2 = 2\pi$, we find:

$$v_{rms} = \sqrt{\frac{A^2}{4\pi}\{[2\pi - 0] - [0 - 0]\}}$$

$$= \sqrt{\frac{A^2}{2}}$$

$$v_{rms} = \frac{A}{\sqrt{2}}$$

This relationship is often expressed to 3 sf in the equivalent form:

$$v_{rms} = 0.707A$$

By integrating over part of the cycle, equation (1) above is used for finding v_{rms} for waveforms such as those of Figure 10.7.

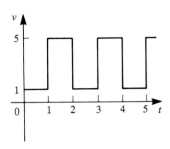

Figure 10.8

Test yourself 10.3

Give all answers to 3 dp where appropriate.

After sketching or examining the graph, find v_{av} for the period stated.

1. $v = 2t + 6$, for $t_1 = 1$ to $t_2 = 2$.
2. $v = t/2 + 3$, for $t_1 = 0$ to $t_2 = 8$.
3. The waveform of Figure 10.8, for **a** $t_1 = 1$ to $t_2 = 3$,
   b $t_1 = 1$ to $t_2 = 5$,
   c $t_1 = 0.5$ to $t_2 = 4.5$.

By integration, find v_{av} and v_{rms} for the period stated.

4. $v = t(3 - t)$, from $t_1 = 1$ to $t_2 = 3$.
5. $v = 2e^t$, from $t_1 = 0$ to $t_2 = 3$.
6. $v = 2 \sin t$, from $t_1 = 3$ to $t_2 = 4$.
7. $v = \sin 3t$, from $t_1 = 0$ to $t_2 = \pi$.
8. An alternating voltage is given by $y = 5 \sin 6\pi t$. Find the mean voltage and the rms voltage over the period from $t_1 = 0$ to $t = 0.05$ s.
9. Calculate the rms voltage of an alternating voltage of amplitude 300 V.
10. Calculate the amplitude of the alternating mains voltage, $v_{rms} = 240$ V.

11 Equations for actions

> Chapter 9 explained how a derivative expresses the rate of change of one variable with respect to another. It expresses action. For this reason, derivatives are useful for expressing the action of almost any kind of system. We use derivatives to describe the actions of anything from a machine such as a diesel engine to a living organism such as a growing wheat seedling, and even including systems such as the economy of a country. We also use them for describing the activities of an electronic circuit.
>
> You need to know about differentiation (Chapter 9) and integration (Chapter 10).

Using a set of equations to describe a system is called **modelling**. The equation or equations are designed so as to behave in the same kind of way in which the system behaves. By inserting different values for the variables of the equation, we can predict how the system will behave under different circumstances. This can save time and money: for example, modelling the behaviour of a jet aeroplane engine using a computer program avoids the need to build the engine and to subject it to countless mechanical modifications before we arrive at the best design. It also makes it possible to predict the effects of actions that would be impracticable or take too long to come to full effect. For example, we could model the effect of doubling Income Tax on the rate of failure of businesses. Modelling also provides us with useful insight into the complications of circuit design.

A simple model

As an example of an extremely simple model, we will write an equation to describe the action taking place in Figure 11.1. A constant current generator is feeding current i into a capacitor. From the definition of the ampere as 1 coulomb per second, we see that the *rate of accumulation of charge* on the capacitor is i coulomb per second. We express this using a derivative to represent the rate of accumulation of charge, in coulomb per second:

$$\frac{dq}{dt} = i$$

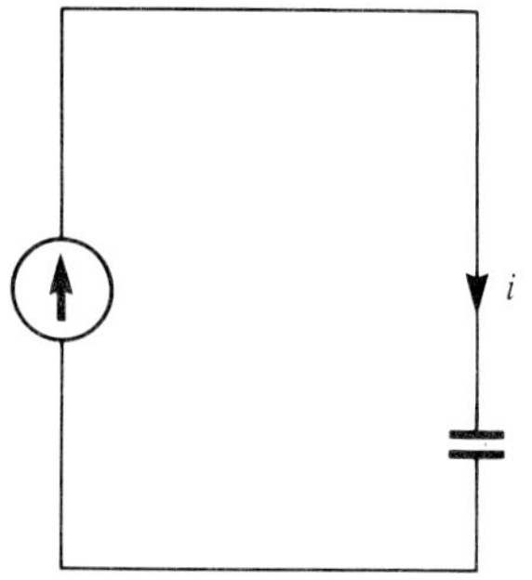

This equation is a *model* of the circuit. Since the equation contains a derivative, or differential, it is called a **differential equation**. Such equations are by far the most useful ones for modelling.

Figure 11.2 illustrates another way of arriving at the same equation. Here we show how accumulated charge varies with time. Starting with an uncharged capacitor, charge builds up steadily as current flows into the capacitor. The charge q at time t is given by:

Figure 11.1

$$q = it$$

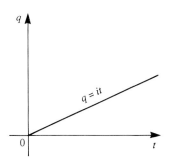

Figure 11.2

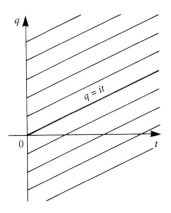

Figure 11.3

Differentiating with respect to t provides the equation for the rate of charge accumulation:

$$\frac{dq}{dt} = i$$

This is the same equation as above. Having written the differential equation which, basically, is just re-stating the definition of the ampere in mathematical terms, the next step is to solve it. The **solution** is an equation in q which has *no* derivative in it. The way to obtain this is to integrate the equation. The first stage is to separate the variables, putting all terms involving q (the dependent variable) on the left and all terms involving t (the independent variable) on the right:

$$dq = i \cdot dt$$

Integrating with respect to q on the left and with respect to t on the right:

$$\int dq = i \int dt$$
$$\Rightarrow \quad q = it + c$$

where c is the constant of integration (page 142). To complete the solution we have to establish a value for c. Figure 11.3 illustrates the fact that the integration has produced a whole **family of curves**, all of which are solutions of the original differential equation. Although only 10 members of the family are drawn in Figure 11.3, in infinite number of such curves exists. All are straight lines, parallel to each other with gradient dq/dt, with their y-intercepts equal to different values of c (compare $y = mx + c$, page 56). The y-intercept can be considered to be the initial charge, if any, present on the capacitor. One of these curves, the one that passes through the origin, is the one from which we derived our differential equation. But it could have been derived from *any one* of the curves in Figure 11.3. What this means in practice is that the rate of accumulation of charge is unaffected by the initial charge on the capacitor.

Solving equations with separable variables

1 Separate the variables (including dx, dy, etc.) to left and right.
2 Integrate both sides.
3 Make the independent variable the subject.

If we are given an additional piece of information, such as the initial charge c, we can calculate the exact charge at any subsequent time. For example, if $c = 2$ V and $i = 3$ A, the charge at time $t = 5$ s is:

$$q = 3 \times 5 + 2 = 17 \, C$$

This is an instance of being able to evaluate an indefinite integral when given an extra piece of information (page 144). When we are modelling, we refer to such extra information as the **boundary conditions**.

Growth and decay

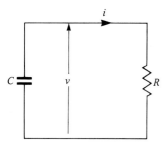

Figure 11.4

The circuit of Figure 11.4 shows a slightly more complicated situation, which requires a slightly more complicated model. A capacitor, capacitance C, is charged to voltage v by connecting it briefly to a power supply and then removing it. It is then connected to a resistor, resistance R as shown. The capacitor discharges through the resistor and, as it does so, v falls. The complication here is that the current i is not constant. It is a maximum when the resistor is first connected because at that time there is the full voltage v to drive the current. But, as current flows and the capacitor becomes partly discharged, v falls and drives the current less strongly. Both v and i fall, more and more slowly, until the capacitor is, for practical purposes, discharged. This is the action to be modelled.

We begin by writing down some of the more obvious relationships found in this circuit, remembering that a small letter, such as q or i refers to an instantaneous value that is probably changing, while capital letters such as R and C refer to fixed values. Here are three relevant equations:

(1) $q = vC$ (charge held is voltage *times* capacitance; the definition of capacitance)
(2) $v = iR$ (Ohm's Law)
(3) $i = dq/dt$ (definition of the ampere)

From (1) and (2):

$$q = vC = iRC$$

Then, using (3)

$$q = iRC = \frac{dq}{dt} \cdot RC$$

From this we obtain the differential equation:

$$\frac{dq}{dt} = \frac{-1}{RC} \cdot q$$

In which $1/RC$ is a constant. The right side of the equation is negated because $\frac{dq}{dt}$ is negative, the amount of charge *decreasing* with time.

This equation is one of very many which have the form:

$$\frac{dx}{dy} = kx$$

Growth and decay equation

In some books this is written as:

$$R \cdot \frac{dq}{dt} + \frac{q}{C} = 0$$

which is just the same as the equation quoted here, with the terms rearranged and multiplied by R.

in which k is a constant. Such equations represent systems in which there is growth or decay. In the example above there is decay of the voltage and current as the capacitor discharges. The equation also models other decays such as radioactive decay, or the cooling of a hot body under constant conditions. It models growth such as that of a plant or of a population, or the increase of an interest-bearing account at a bank.

Solving the decay equation

Begin with the equation:

$$\frac{dq}{dt} = \frac{-1}{RC} \cdot q$$

Separate the variables:

$$\frac{dq}{-q} = \frac{dt}{RC}$$

Integrate both sides:

$$\int \frac{1}{-q} \, dq = \frac{1}{RC} \int dt$$

$$\Rightarrow \quad \ln \left| \frac{A}{-q} \right| = \frac{t}{RC}$$

$$\Rightarrow \quad \ln \left| \frac{A}{q} \right| = \frac{t}{RC}$$

Note that the log includes the *constant of integration* A, as explained on page 144, and that we take the *absolute* value of A/q as explained on page 147.

Using the relationship between logs and exponentials (page 26) this equation becomes:

$$\frac{A}{q} = e^{t/RC}$$

Taking reciprocals (page 36):

$$\frac{q}{A} = e^{-t/RC}$$

$$\Rightarrow \quad \mathbf{q = A e^{-t/RC}}$$

This solution of the differential equation does not provide the value of the constant of integration. It represents a family of equations with different values of A. Some of these are shown in Figure 11.5.

To evaluate A we need further information. If we are told that discharging begins with $t = 0$, substituting in the solution gives:

$$q = Ae^0 = A$$

The boundary condition is that A is the initial charge on the capacitor.

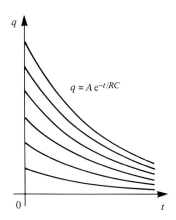

Figure 11.5

Example

A capacitor is charged to 10 V. It is then discharged through a resistor to 2 V in 3 s. If its capacitance is 10 µF, what is the value of the resistor?

Calculating the charges on a 10 µF capacitor corresponding to p.d.s. of 10 V and 2 V, we find that q drops from 100 µC to 20 µC in 3 s.

At the start, $t = 0$; substituting values of q and t:

$$100 \times 10^{-6} = Ae^0$$

$\Rightarrow \qquad A = 100 \times 10^{-6}$

After 3 s, $t = 3$; substituting values of t, A, and C:

$$20 \times 10^{-6} = 100 \times 10^{-6} \times e^{-3/RC}$$

$\Rightarrow \qquad 0.2 = e^{-3/RC}$

$\Rightarrow \qquad \ln(0.2) = \dfrac{-3}{RC}$

$\Rightarrow \qquad R = \dfrac{-3}{\ln(0.2) \times C} = \dfrac{-3}{-1.6094 \times 10 \times 10^{-6}}$

$\Rightarrow \qquad R = 186\,400$

The resistance is 186 kΩ (3 sf).

Differential equations of growth and decay all have the same form, so can all be solved by the method given above. Having worked the solution step-by-step once, there is no need to repeat the steps every time. We can simply write out the solution using the rules set out in the box.

Identifying and solving growth and decay equations

1 Make dy/dx the subject. The form should then be:

   $dy/dx = ky$

2 Rewrite in the form:

   $y = Ae^{kx}$

3 If given boundary conditions, substitute known values and solve for unknown values.

Test yourself 11.1

1 Solve the equation $dy/dx = 3$. Given that $y = 7$ when $x = 2$, find y when $x = 9$.

2 Solve the equation $dy/dx = 5$. Given that $y = 12$ when $x = 2$, find y when $x = 1$.

3 Solve the equation $dy/dx = 3y$. Given that $y = 1$ when $x = 0$, find y when $x = 3$.

4 Solve the equation $dy/dx + 4y = 0$. Given that $y = 3$ when $x = 1$, find y when $x = 0.5$

5 A 200 μF capacitor is charged to 5 V and then discharged through a 100 kΩ resistor, as in Figure 11.4. How long does it take to discharge the capacitor (a) to 1 V (b) to 0.5 V?

6 A 1 kΩ resistor is connected in series with a 2 H inductor. The current falls to 2 A in 0.5 ms. Given that $i = Ae^{-Rt/L}$, where L is the inductance, find (a) the initial current, (b) how long the current takes to fall to 1 A.

First-order equations

A first-order linear differential equation is one which includes only the first differential, not second or higher differentials. The basic equation is:

$$\frac{dy}{dx} + f(x)y = g(x)$$

$f(x)$ and $g(x)$ are two functions of x. By making $f(x)$ a constant and $g(x)$ equal to zero, we get back to the growth and decay equation (page 160) so this, too, is a first-order equation.

We will model an electronic circuit by using the first-order equation, and then show how to solve it.

Figure 11.6 gives the circuit, in which a voltage v is applied to a resistor and inductor wired in series. The first task is to find an equation to model this circuit. This is done, as usual, by writing down various well-known relationships and then trying to combine some of them into an equation.

The voltage could be constant but we will assume that it is varying. At any given instant, the voltage across the resistor is v_R, and the voltage across the inductor is v_L. By KVL:

$$v_L + v_R = v \quad (1)$$

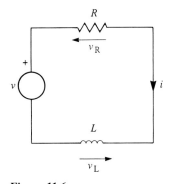

Figure 11.6

Other relevant equations are:

$$v_R = Ri \quad \text{(Ohm's Law)} \quad (2)$$

$$v_L = L\,di/dt \quad (3)$$

Substituting (2) and (3) in (1):

$$L\frac{di}{dt} + Ri = v$$

Dividing both sides by L so that the coefficient of the first term is unity:

$$\frac{di}{dt} + \frac{R}{L} \cdot i = \frac{v}{L}$$

This is a **first-order differential equation** of the form given above, with $f(t) = R/L$ and $g(t) = 1/L$. Both functions in this example are constants, not involving i.

The discussion which follows has several stages, all of which are based on the differentiating and integrating techniques that you have already studied.

Follow through the discussion, step-by-step, but there is no need to memorize it. See it as an example of the way a differential equation is solved. Later we will give a few short-cut rules for solving equations of this type.

The first stage in solving equations of this type is to find an **integrating factor**. This is a factor which, when used to multiply both sides of the equation, leads to another equation which is simpler to solve. The integrating factor is found this way:

In this example

- Take the coefficient of the term in i; $\dfrac{R}{L}$

- Find its integral with respect to t; $\displaystyle\int \dfrac{R}{L}\, dt = \dfrac{Rt}{L}$

- Use the result as a power of e. $e^{Rt/L}$

Multiply all terms in the equation by the integrating factor, $e^{Rt/L}$:

$$e^{Rt/L}\dfrac{di}{dt} + \dfrac{R}{L}e^{Rt/L}\cdot i = \dfrac{v}{L}e^{Rt/L}$$

In the next step we replace the two terms on the left by a single term. The box shows the reasoning behind this; the single term, the derivative of $e^{Rt/L}i$, gives the two terms which it now replaces.

$$\dfrac{d}{dt}(e^{Rt/L}\, i) = \dfrac{v}{L}e^{Rt/L}$$

Derivative of a product

See page 130.

Let $u = e^{Rt/L}$ and $v = i$

Differentiating:

$$\dfrac{du}{dt} = \dfrac{R}{L}e^{Rt/L} \quad \text{and} \quad \dfrac{dv}{dt} = \dfrac{di}{dt}$$

$$\Rightarrow \quad \dfrac{d}{dt}(e^{Rt/L}\, i) = \dfrac{d(uv)}{dt}$$

$$= u\dfrac{dv}{dt} + v\dfrac{du}{dt}$$

$$= e^{Rt/L}\dfrac{di}{dt} + \dfrac{i}{L}Re^{Rt/L}$$

Integrating both sides of the equation with respect to t (on the left, integrating a derivative restores the original term):

$$e^{Rt/L}\, i = \int \dfrac{v}{L} e^{Rt/L}\, dt + A$$

A is the constant of integration. Divide both sides by $e^{Rt/L}$:
This is the solution of the equation. Note the two instances where e has a negative index.

$$i = e^{-Rt/L} \int \frac{v}{L} e^{Rt/L} \, dt + Ae^{-Rt/L} \tag{4}$$

Given values of R, L, and an equation which gives v as a function of t, the value of i can be calculated at any instant of time. However, as usual, it is necessary to know the border conditions in order to calculate the value of A.

Equation (4) is used as it stands, or it is simplified by specifying certain conditions. For example, if v is known to be constant, the first term on the right side of the equation becomes v/R (see box), and the equation reduces to:

$$i = \frac{v}{R} + Ae^{-Rt/L} \tag{5}$$

The integration in detail

If v is constant, v/L can be placed before the integral sign, giving:

$$e^{-Rt/L} \frac{v}{L} \int e^{Rt/L} \, dt$$

The integrand is of the form e^{ax}, with $a = R/L$. The integral of e^{ax} is $\frac{1}{a} e^{ax}$ (page 143). Thus integrating the expression above gives:

$$e^{-Rt/L} \frac{v}{L} \cdot \frac{L}{R} e^{Rt/L}$$

Multiplying the powers of e (adding their indices) gives $e^0 = 1$; also cancelling the Ls reduces the expression to:

$$\frac{v}{R}$$

If we stipulate a border condition, that i has the value i_0 when $t = 0$, we can find A by substituting in (5):

$$i_0 = \frac{v}{R} + Ae^0 = \frac{v}{R} + A$$

$$\Rightarrow \quad A = i_0 - \frac{v}{R}$$

Substituting this in equation (5):

$$i = \frac{v}{R} + \left(i_0 - \frac{v}{R}\right) e^{-Rt/L} \tag{6}$$

Figure 11.7 is a graph of this equation, when $i_0 = 0.01$ A, $v = 5$ V, $R = 100\,\Omega$, and $L = 100$ mH. The current rises from its initial value, gradually approaching but never quite reaching a steady state with $i = v/R$. In the steady

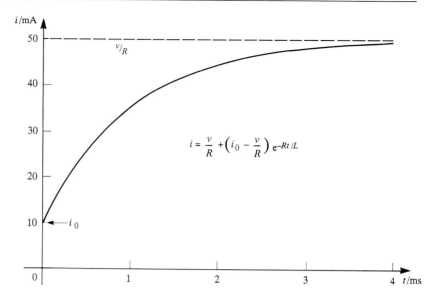

Figure 11.7

Solving first-order linear differential equations

Example

To solve $\dfrac{dy}{dx} + f(x)y = g(x)$ $\dfrac{dy}{dx} - y = 2e^x$

1. Identify $f(x)$ and $g(x)$ $f(x) = -1;\ g(x) = 2e^x$
2. Find $F(x) = \int f(x)\,dx$ $F(x) = \int -1\,dx = -x$
3. The integrating factor is $e^{F(x)}$ Integrating factor is e^{-x}
4. Find $G(x) = \int g(x)\,e^{F(x)}\,dx$ $G(x) = \int 2e^x\,e^{-x}\,dx$
                                            $\quad\quad\, = \int 2\,dx = 2x$
5. Then $y = e^{-F(x)}\,G(x) + Ae^{-F(x)}$ $y = 2xe^x + Ae^x$

Where A is the constant of integration, found by using boundary conditions.

If $f(x)$ is constant and $g(x) = 0$,

we have the growth and decay equation:

To solve $\dfrac{dy}{dx} + f(x)y = 0$ $\dfrac{dy}{dx} + 3y = 0$

1 and 2, as above $f(x) = 3;\ g(x) = 0$
                     $F(x) = \int 3\,dx = 3x$

3. Then $y = Ae^{-F(x)}$ $y = Ae^{-3x}$

4. Or solve by separating variables, as on page 159.

state there is no change of voltage or current, so the inductive effect disappears and the current is limited only by the resistor, plus the low resistance of the inductor coil, which we can assume is included with R.

If the circuit had been in a steady state *before* $t = 0$, the current i_0 may be taken to be the current of a previous steady state. In this state, $v = i_0 R = 0.01 \times 100 = 1$ V. In this way, the graph illustrates the **transient** state, when the voltage is *instantly* increased from 1 V to 5 V. Equations of this kind are useful in examining transient conditions in circuits.

More modelling

The operation of the model may be explored further, to discover what happens if v is a varying voltage instead of being constant. As an example, v might vary linearly according to the equation:

$$v = v_0 + mt$$

where v_0 is the initial voltage at time $t = 0$ and increases steadily at m volts per second. This function for v is substituted for the instantaneous value v in equation (4):

$$i = e^{-Rt/L} \int \frac{v_0 + mt}{L} e^{Rt/L} \, dt + A e^{-Rt/L} \qquad (7)$$

The integral now includes *two* terms involving t, so the integration is not as straightforward as it was before. Integrating and finding the value of A leads to this solution:

$$i = \frac{v_0 R + m(Rt - L)}{R^2} + \left(i_0 - \frac{v_0 R - mL}{R^2}\right) e^{-Rt/L} \qquad (8)$$

The details of the integration are left to the enthusiastic reader. It can all be done with the techniques described in Chapters 9 and 10, including integration by parts (page 150).

The solution of equation (8) can be simplified in various ways:

- if we make $m = 0$, simulating constant voltage, the solution becomes identical with equation (6);
- if we make $L = 0$, which removes the inductor from the circuit, the solution simplifies to:

$$i = \frac{v_0 + mt}{R} \qquad (9)$$

This is a version of Ohm's Law, with varying voltage, and no inductive action. These simplifications give the sort of result expected, indicating that the rather complicated equation is basically correct.

Figure 11.8 shows the effects of inserting different values of m into the solution with $L = 100$ mH, $R = 100\,\Omega$ and $v_0 = 5$ V. We still have the initial current $i_0 = 100$ mA. The dashed lines represent equation (9), showing what i would be in the absence of the inductor. With $m = 100$, i decreases rapidly from 100 mA, swinging round during the transient state until it runs parallel and below the dashed line in the steady state. It never reaches the dashed line because v and therefore i is increasing, inducing a voltage across the inductor.

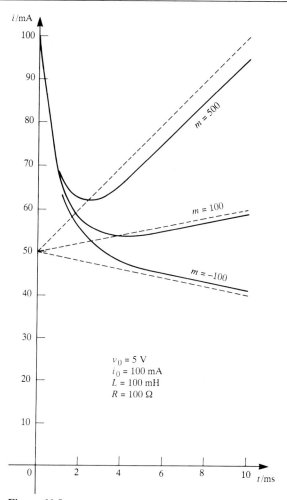

Figure 11.8

This is the stage at which the second term in equation (8) has become so small that it can be ignored and we have:

$$i = \frac{v_0 R + m(Rt - L)}{R^2} \qquad (10)$$

The distance of the $m = 100$ line below the dashed line is the current due to the induced voltage. It can be calculated from differences in the value of i in equation (9) and (10):

$$\begin{aligned} i_{\text{diff}} &= \frac{v_0 + mt}{R} - \frac{v_0 R + m(Rt - L)}{R^2} \\ &= \frac{v_0 R + mRt - v_0 R - mRt + mL}{R^2} \\ &= \frac{mL}{R^2} \end{aligned}$$

Substituting $m = 100$, $L = 10\,\text{mH}$ and $R = 100\,\Omega$ into this equation, we find $i_{\text{diff}} = 1\,\text{mA}$, as is shown on the graph. The result is positive, indicating that the current without the inductor is greater than the current with the inductor.

If $m = 500$, the gap between the lines is wider owing to the increased rate of change of i inducing a larger voltage in the inductor. Substituting in the equation above, we find that it is $5\,\text{mA}$. Once again, the model equation shows the features that would be found in a practical circuit.

With $m = -100$ the current approaches the dashed line and finally runs parallel with it, with $1\,\text{mA}$ above, as the induced current is in the opposite direction now. Note the symmetry about the $50\,\text{mA}$ level, of the steady-state parts of the curves for $m = 100$ and $m = -100$.

The effects of different inductors are shown in Figure 11.9. The initial current is taken as zero in this series, so that we see the transients immediately following switching on the power. All curves correspond to $m = 100$. The curve for $L = 100\,\text{mH}$ rises and runs below the dashed line, with a gap representing $1\,\text{mA}$ due to induction. Reducing L to $40\,\text{mH}$ gives the circuit a much more rapid response and the final current approaches the dashed line more closely. With a very large inductor, $L = 250\,\text{mH}$, there is considerable damping. During the period of time shown on the graph, the final current only just begins to rise parallel with the dashed line. The equation behaves just as we would expect a real circuit to behave.

The data for the curves in Figures 11.7 to 11.9 was calculated in a few minutes using a 6-line computer program written in BASIC. This is a quicker and more precise procedure than actually building the circuit and monitoring

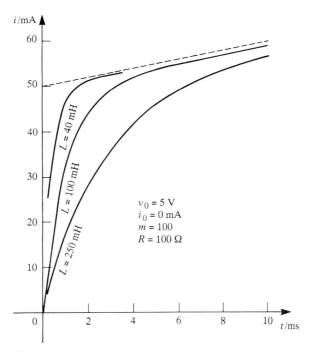

Figure 11.9

its behaviour with instruments. The model allows component values to be changed and their effects observed until the final circuit design is arrived at.

Test yourself 11.2

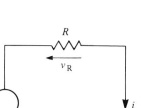

Figure 11.10

1 Given the circuit of Figure 11.6 and equation (8): (a) write out a simplified version of the equation when $v_0 = 0$ and $i_0 = 0$. (b) If $m = 100$ Vs^{-1}, $L = 50$ mH, $R = 250\,\Omega$ and the border conditions are as in part (a), calculate the current flowing after $80\,\mu s$ (in μA, to 2 dp).

2 Use the worked example of Figure 11.6 as a model for solving this question. Unless you are certain, check the answer at each stage before proceeding with the next.
  In the circuit of Figure 11.10, the voltages at any instant are:
  v (input voltage), $v_c = q/C$, and $v_R = R \cdot \dfrac{dq}{dt}$.

  a Write an equation relating these voltages.
  b Rewrite the equation of part a as a differential equation.
  c Calculate the integration factor.
  d Solve the equation, to obtain an equation for q.
  e Assuming that v is constant, simplify the equation of part d.
  f Assuming the border conditions, $q_0 = 0$ when $t = 0$, calculate A, and write an equation for the value of q at any given time t.
  g Use the equation of part f to obtain an equation for v_c at any given time t.
  h If $v = 6$ V, $C = 150$ nF, $R = 10$ kΩ, calculate v_c when $t = 0.5$ ms.

Solve the following differential equations.

3 $\dfrac{dy}{dx} + 4y = 0$ 4 $\dfrac{dy}{dx} - by = 0$ (b is a constant)

5 $\dfrac{dy}{dx} + \dfrac{y}{x} = 0$ 6 $\dfrac{dy}{dx} + 2y = x$ [Integrate by parts to find $G(x)$.]

7 $\dfrac{dy}{dx} + 4y = e^x$ 8 $\dfrac{dy}{dx} + y = x^2 e^{-x}$

Second-order equations

The general second-order differential equation has the form:

$$\dfrac{d^2y}{dx^2} + f(x)\dfrac{dy}{dx} + g(x)y = h(x) \tag{1}$$

where $f(x)$, $g(x)$ and $h(x)$ are all functions of x. As might be expected, working with second-order equations can become very much more complicated than solving first-order equations, so we shall limit the discussion to second-order equations of a relatively simple kind. We will make $f(x)$ and $g(x)$ constants, and make $h(x)$ equal to zero. This reduces equation (1) to:

$$\frac{d^2y}{dx^2} + f\frac{dy}{dx} + gy = 0 \qquad (2)$$

where f and g are constants.

To solve this equation we adopt the strategy of trying to guess the answer and then checking it to see if it gives a sensible result. A sensible guess is that one solution of the equation is:

$$y = e^{mx}$$

where m is a constant. This conveniently is one of the standard forms. Basing the next step on this assumption, we differentiate it twice to find:

$$\frac{dy}{dx} = me^{mx} \quad \text{and} \quad \frac{d^2y}{dx^2} = m^2 e^{mx}$$

Substituting these values in equation (2):

$$m^2 e^{mx} + fme^{mx} + ge^{mx} = 0$$
$$\Rightarrow \quad e^{mx}(m^2 + fm + g) = 0$$
$$\Rightarrow \quad m^2 + fm + g = 0 \qquad (3)$$

This equation is known as the **auxiliary equation** of the differential equation. If we can find the values of m which satisfy this equation we shall be able to write out an expression for y, a solution of the differential equation.

Solving equation (3) for m is easy, because it is a quadratic equation, and we have already seen how to solve these (page 39). If the values of f and g are suitable it is possible to solve it by factorizing but, to cover all cases, we will use the quadratic formula (page 40). The corresponding coefficients are:

$$a = 1 \quad b = f \quad c = g$$

There are three possible kinds of solution, depending on whether the value D of the discriminant (page 40) is positive, zero or negative. So the next step is to calculate:

$$D = f^2 - 4g$$

We consider the three possible cases in turn.

Solving with positive discriminant

The auxiliary equation has two real roots:

$$m = \frac{-f \pm \sqrt{D}}{2}$$

Call the two roots m_1 (using the positive square root) and m_2 (using the negative square root). Having obtained the roots, we write out the solution to the differential equation, which has the form:

$$y = Ae^{m_1 x} + Be^{m_2 x} \qquad (4)$$

This equation contains two constants A and B. Like the constant A in the solutions on pages 162 and 165, the values of these can be found if we are told the boundary conditions. The difference here is that there are *two* constants and we need to be given *two* pieces of information. Often we are told the initial value of y and also the initial rate at which it is changing (dy/dx).

Before going on to consider the next case, we should justify the introduction of equation (4) without any explanation of how it is derived. The simplest course is to test whether or not it satisfies the differential equation, equation (2). Take an example:

Solve $\dfrac{d^2y}{dx^2} - 3\dfrac{dy}{dx} + 2y = 0$ (5)

Here $f = -3$ and $g = 2$

$\Rightarrow \qquad D = (-3)^2 - 4 \times 2 = 1$

This is greater than zero, so there are two real roots. Calculate roots:

$$m_1 = \frac{3+1}{2} = 2 \qquad m_2 = \frac{3-1}{2} = 1$$

Write out the solution:

$y = Ae^{2x} + Be^x$ (6)

Now to check that equation (6) *really is* a solution to equation (5). Given the value of y from equation (6), differentiate this to find:

$\dfrac{dy}{dx} = 2Ae^{2x} + Be^x$ (7)

Differentiate again to find:

$\dfrac{d^2y}{dx^2} = 4Ae^{2x} + Be^x$ (8)

We now have values for y, dy/dx and d^2y/dx^2 to substitute in equation (5):

$$\frac{d^2y}{dx^2} - 3\frac{dy}{dx} + 2y = (4Ae^{2x} + Be^x) - 3(2Ae^{2x} + Be^x) + 2(Ae^{2x} + Be^x)$$

$$= Ae^{2x}(4 - 6 + 2) + Be^x(1 - 3 + 2)$$

$$= 0$$

The substituted values equate to zero, as does the original version of equation (5), so showing that equation (6) is a solution.

Solving with zero discriminant

The auxiliary equation has a double root, $m = -f/2$. In this case the solution is written out as:

$y = Ae^{mx} + Bxe^{mx}$ (9)

As above, the solution has been presented without proof, but an example will show that it is justified.

Solve $\dfrac{d^2y}{dx^2} + 2\dfrac{dy}{dx} + y = 0$ (10)

$f = 2$ and $g = 1$.

$\Rightarrow \quad D = (2)^2 - 4 \times 1 = 0$

The discriminant is zero, there is a double root.
Calculate $m = -2/2 = -1$.
Write the solution:

$y = Ae^{-x} + Bxe^{-x}$ (11)

To prove the validity of the solution, differentiate equation (11), using the Product Rule (page 130) for the second expression:

$$\dfrac{dy}{dx} = -Ae^{-x} + B(-xe^{-x} + e^{-x})$$

$$= -Ae^{-x} - Bxe^{-x} + Be^{-x}$$

Differentiate again, using the Product Rule for the second expression:

$$\dfrac{d^2y}{dx^2} = Ae^{-x} - B(-xe^{-x} + e^{-x}) - Be^{-x}$$

$$= Ae^{-x} + Bxe^{-x} - 2Be^{-x}$$

Substitute these values in equation (10):

$$\dfrac{d^2y}{dx^2} + 2\dfrac{dy}{dx} + y = (Ae^{-x} + Bxe^{-x} - 2Be^{-x}) + 2(-Ae^{-x} - Bxe^{-x} + Be^{-x})$$
$$+ (Ae^{-x} + Bxe^{-x})$$
$$= Ae^{-x}(1 - 2 + 1) + Bxe^{-x}(1 - 2 + 1) + Be^{-x}(-2 + 2)$$
$$= 0$$

This demonstrates that equation (11) is a solution of equation (10).

Solving with negative discriminant

The auxiliary equation has two complex roots (page 248). Obtaining a solution is slightly more difficult because of this. Having calculated D, find:

$$k = \dfrac{\sqrt{-D}}{2}$$

taking the positive square root. Now write out the solution:

$y = Ae^{-fx/2} \cos kx + Be^{-fx/2} \sin kx$

As with the solutions of the other cases, there are two constants to be evaluated using two given border conditions. It can be shown by differentiating twice and substituting, as above, that this solution satisfies the differential equation, but the calculations are lengthy and we will take this for granted. An example of an equation with complex roots is given below.

A second-order model

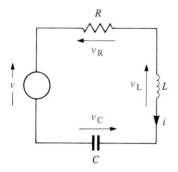

Figure 11.11

In Figure 11.11 a voltage v is applied to a circuit consisting of a resistor, an inductor and a capacitor wired in series. At any instant, by KVL:

$$v_L + v_R + v_C = v$$

For each of the terms on the left we substitute others based on the laws of electronics:

$$L\frac{di}{dt} + Ri + \frac{q}{C} = v$$

where i is the instantaneous current and q is the charge on the capacitor. Simplifying the situation by assuming that v is constant, we differentiate both sides of the equation with respect to t:

$$L\frac{d^2i}{dt^2} + R\frac{di}{dt} + \frac{i}{C} = 0$$

In differentiating the third term we rely on the fact that, by definition, $dq/dt = i$. Now divide throughout by L:

$$\frac{d^2i}{dt^2} + \frac{R}{L} \cdot \frac{di}{dt} + \frac{1}{LC} \cdot i = 0 \tag{12}$$

This is a second-order linear differential equation of the type described on page 171. In this equation, $f = R/L$ and $g = 1/LC$ and the auxiliary equation is:

$$m^2 + \frac{R}{L}m + \frac{1}{LC} = 0$$

Solving the auxiliary equation, we calculate the discriminant:

$$D = f^2 - 4g = \frac{R^2}{L^2} - \frac{4}{LC}$$

Whether or not this is positive, zero or negative depends on the values of the components of the circuit. We will look at two of the many possibilities:

First set of values: With $R = 1\,\text{k}\Omega$, $L = 100\,\text{mH}$ and $C = 420\,\text{nF}$, we find:

$$D = 4.76 \times 10^6$$

This is positive so the equation has two real roots:

$$m = \frac{-f \pm \sqrt{D}}{2} = \frac{-R/L \pm \sqrt{D}}{2}$$

$$m_1 = \frac{-10\,000 + 2182}{2} = -3909$$

$$m_2 = \frac{-10\,000 - 2182}{2} = -6091$$

Thus the solution to the equation is:

$$i = Ae^{-3909t} + Be^{-6091t} \tag{13}$$

Do not worry about the apparently large indices of e; remember that t is likely to be only a few microseconds, reducing the indices to reasonable amounts.

This solution has two unknown constants, A and B, often referred to as **arbitrary constants**. Since there are many possible values that A and B may take, depending on the border conditions, equation (13) is called the **general solution**. In order to find A and B, we need *two* pieces of information about boundary conditions. For one of them, suppose that the zero current is flowing at time $t = 0$. Substituting $t = 0$ and $i = 0$ in equation (13):

$$0 = A + B$$

$$B = -A$$

Equation (13) may now be simplified to:

$$i = A(e^{-3909t} - e^{-6091t}) \tag{14}$$

Next differentiate equation (14):

$$\frac{di}{dt} = A(-3909e^{-3909t} + 6091e^{-6091t})$$

This is the rate of change of current, which we might be given as the second border condition. Assume it is $0.1\,A\,s^{-1}$ when $t = 0$. Substituting this value:

$$A = \frac{0.1}{6091 - 3909}$$

$$= 4.58 \times 10^{-5}$$

Substituting this in equation (14) gives a solution to equation (13) with this particular combination of components and border conditions:

$$\boldsymbol{i = 4.58 \times 10^{-5}\,(e^{-3909t} - e^{-6091t})} \tag{15}$$

We call this a **particular solution**. Figure 11.12 shows the graph of this. The circuit has previously had a voltage applied to it to produce a certain amount

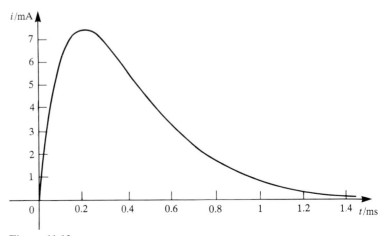

Figure 11.12

of charge on the capacitor and a magnetic field in the inductor. Then, at time $t = 0$, when (as it happens) no current is flowing and the current is increasing (from some negative value) at 0.1 A s^{-1}, the voltage is suddenly held constant. Obviously, with no further change in applied voltage, the current eventually falls to zero. The graph shows the behaviour of the current during this transition period. It shows that the current rises from zero to approximately 1 µA during the first 0.01 ms. This is equivalent to 0.1 A s^{-1} as specified for the second boundary condition. But, from the very beginning, the capacitor and inductor influence the current very strongly. The rate of increase is not maintained. After about 0.2 ms the current is no longer increasing but is falling toward zero. It falls rapidly at first and then more and more slowly. After about 1.5 ms there is virtually no current. The graph demonstrates that the current is strongly damped. Since it is returned to zero relatively rapidly, without ever reversing in direction, we say that it is **over-damped**. Thus equation (15) is a model of an over-damped LCR circuit.

Second set of values: For this example we change only the value of the capacitor, reducing it from 420 nF to 10 nF. This alteration might be expected to reduce the damping. Let us examine the result of this change. First we re-calculate D, which is now -39×10^8. The main point of interest is that D is now negative, and therefore the auxiliary equation has complex roots (page 173). The next step is to calculate k:

$$4k^2 = -D = 3.9 \times 10^9$$

$$\Rightarrow \quad k = 31285$$

Writing out the general solution, as on page 173, with $f/2 = R/2L = 5000$:

$$i = Ae^{-5000t} \cos 31225t + Be^{-5000t} \sin 31225t \qquad (16)$$

Once again there are large numbers in the equation, but t will be small, reducing the indices and coefficients to manageable values. If, as before, $i = 0$ when $t = 0$, the term containing B disappears, because $\sin 0 = 0$. But $\cos 0 = 1$, so equation (16) gives:

$$0 = A$$

This means that the term with A is eliminated from equation (16) and the solution becomes:

$$i = Be^{-5000t} \sin 31225t \qquad (17)$$

Now to calculate B. In order to use the information of the rate of change of current, di/dt, we differentiate equation (17), using the Product Rule:

$$\frac{di}{dt} = B(e^{-5000t} \, 31225 \cos 31225t - 5000 e^{-5000t} \sin 31225t) \qquad (18)$$

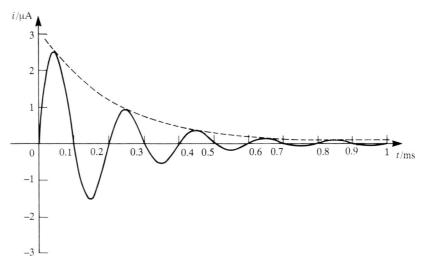

Figure 11.13

When $t = 0$, we will make $di/dt = 0.1$, as in the previous example, so that the only difference between the two circuits is the capacitance. Since $\sin 0 = 0$ and $\cos 0 = 1$, equation (18) reduces to:

$$0.1 = B \times 31\,225$$
$$\Rightarrow \quad B\ 3.20 \times 10^{-6}$$

The particular solution under the given conditions is:

$$i = 3.20 \times 10^{-6} \times e^{-5000t} \times \sin 31225t$$

Figure 11.13 is the graph of this solution. From the equation, we can see that i is the product of two functions, one exponential and the other a periodic sine function. Thus we would expect the graph to be a sine wave with its amplitude decreasing exponentially, and this is just what Figure 11.13 shows. The dashed line joining the peaks of the sine wave falls in an exponential manner. The period of the wave can be seen to be very close to 0.2 ms. This follows from the equation since, with $t = 0.2$ ms, the value of $31\,225t$ is very close to 2π. In radians, this represents one complete cycle. Also the current reaches $1\,\mu A$ during the first 0.01 ms, confirming its initial rate of increase of 0.1 A s^{-1}. The shape of the graph shows how the current alternates with ever-decreasing amplitude, reversing in direction several times until it virtually dies out after about 1 ms. This is the result of reducing the value of the capacitor and hence its damping effect. The circuit is said to be **under-damped**.

Solving second-order equations

With $f(x)$ and $g(x)$ constant and $h(x) = 0$

$$\frac{d^2y}{dx^2} + f\frac{dy}{dx} + gy = 0$$

Calculate $D = f^2 - 4g$

If D positive

Calculate $\left. \begin{array}{l} m_1 = \dfrac{-f + \sqrt{D}}{2} \\[2mm] m_2 = \dfrac{-f - \sqrt{D}}{2} \end{array} \right\} \Rightarrow y = Ae^{m_1 x} + Be^{m_2 x}$

If $D = 0$

Calculate $m = \dfrac{-f}{2} \Rightarrow y = Ae^{mx} + Bxe^{mx}$

If D negative

Calculate $k = \sqrt{\dfrac{-D}{2}}$

$\Rightarrow y = Ae^{-fx/2} \cos kx + Be^{-fx/2} \sin kx$

Test yourself 11.3

Find the general solution for each of the following equations.

1 $\dfrac{d^2y}{dx^2} - 6\dfrac{dy}{dx} + 5y = 0$

2 $\dfrac{d^2y}{dx^2} + \dfrac{dy}{dx} - 12y = 0$

3 $\dfrac{d^2y}{dx^2} - 2\dfrac{dy}{dx} + 5y = 0$

4 $\dfrac{d^2y}{dx^2} + 4\dfrac{dy}{dx} + 4y = 0$

5 $\dfrac{d^2y}{dx^2} + 5\dfrac{dy}{dx} = 0$

6 Find the general solution of this equation:

$\dfrac{d^2y}{dx^2} - 3\dfrac{dy}{dx} + 10y = 0$

If, when $x = 0$, $y = 0$ and $dy/dx = 7$, calculate the values of A and B and write out the particular solution.

7 Find the general solution of this equation:

$\dfrac{d^2y}{dx^2} - 3\dfrac{dy}{dx} + 2y = 0$

If, when $x = 0$, $y = 1$ and $dy/dx = -2$, calculate the values of A and B and write out the particular solution.

8 Find the general solution of this equation:

$\dfrac{d^2y}{dx^2} - 8\dfrac{dy}{dx} + 25y = 0$

If, when $x = 0$, $y = 2$ and $dy/dx = 5$, calculate the values of A and B and write out the particular solution.

9 For the circuit of Figure 11.11, find the particular solution when $L = 100$ mH, $R = 1$ kΩ, $C = 500$ nF. At $t = 0$, the boundary conditions are $i = 5$ µA, and $di/dt = 0.01$ A s^{-1}. Calculate the value of i at $t = 300$ µ (in µA, 3 sf).

10 In Figure 11.11 given $L = 100$ mH, $C = 420$ nF and with border conditions $i = 1$, $di/dt = 0$, calculate the value of R which would make the discriminant of the auxiliary equation equal to zero. Write the particular solution of the equation.

11 With $D = 0$, the circuit of the previous question is said to be *critically* damped. Using the particular solution draw the graph of the current passing in the circuit during the first 1.5 ms after the applied voltage has been made constant.

12 Practical differentials

> Mathematical modelling with differential equations is an essential aid to circuit design. This chapter explains three techniques which make the modelling easier.
> You need to know about partial fractions (page 18), limits (Chapter 8) and differential equations (Chapter 11).

The circuits in the previous chapter are helpful examples of how to build a model with differential equations, but we make several assumptions in order to keep the equations relatively simple. It could be said that, even so, the solutions to the equations are complicated enough. Two crucial assumptions are that the functions of x are either constant or zero. In real-life circuits, neither of these assumptions may be valid. Furthermore, a model may also require differentials of the third or higher orders. On top of that, it often happens that an apparently straightforward expression is very difficult, or even impossible to integrate. The standard routines of integration leave us floundering with pages of symbols and with no valid solution at the end of it.

In this chapter we outline some techniques for solving differential equations without having to integrate difficult expressions.

Euler's method

This is the simplest to do and the easiest to understand, though it lacks the precision of the other methods. However, it is adequate for many practical purposes. We will show how to use this for first-order equations. These have the general form:

$$\frac{dy}{dx} = f(x, y)$$

where $f(x,y)$ may contain terms in x or y or both, and may also include constants.

On page 159 we saw how integrating a differential equation gives rise to a whole family of curves, each one of them being a possible solution to the equation. We select a particular one by specifying the border conditions, which determine a particular point through which the curve to be selected must pass. In Figure 12.1 the curves are just a few of the family produced from a given differential equation and A is the point determined by boundary conditions. The curve drawn in a heavy line is thus the one which is the particular solution to the equation. A problem arises when we are given the differential equation, and the boundary conditions, but it is mathematically too difficult to calculate the equation for the curve by normal integration. It is as if we had a blank area with point A marked and we need to know where the curve goes from there.

PRACTICAL DIFFERENTIALS 181

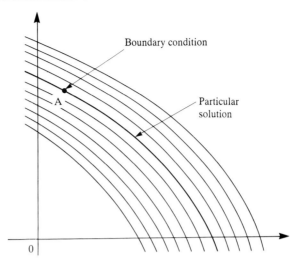

Figure 12.1

One thing we do know is the position of point A. The other thing we know is the gradient of the curve (dy/dx) at this point, because this can be calculated by substituting the known coordinates of point A in the equation. Although we cannot plot the *curve* which passes through point A, we can do the next best thing and plot the *tangent* to the curve at that point, a line with gradient dy/dx. Provided we plot only a *short* tangent, we shall not stray too far from the curve. This takes us to point B (Figure 12.2). Because we have plotted a

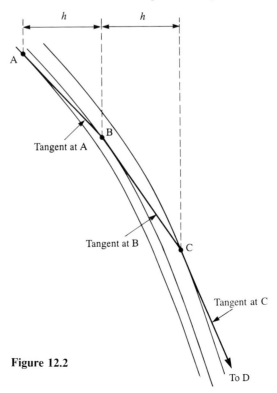

Figure 12.2

tangent, not the curve itself, point B does not actually lie on the same curve as A. Instead it lies on another curve which, provided that we have not moved too far from A, is a very close member of the family.

At B, we repeat the construction, plotting a tangent (to the next curve) by using the coordinates of point B. This gives us the position of point C. Point C is a on a family curve a little further from the original curve but still close enough to be acceptable as an approximation. In this way we proceed to plot as many points as we need and join them together to make an approximation to the curve on which point A lies. As can be seen in Figure 12.2, the final point is *not* on the same curve as point A, giving rise to an error. The error is minimized by drawing shorter tangents at each point. But this means that we can plot only a short part of the curve. Plotting shorter tangents but more of them increases the error again because of the rounding of values which takes place at each stage.

As an example of Euler's method take the equation on page 163:

$$\frac{di}{dt} + \frac{R}{L}i = \frac{v}{L} \tag{1}$$

Here we are using i as the dependent variable instead of y, and t as the independent variable instead of x. It is given that $v = 5$, $R = 100$, and $L = 100\,\text{mH}$ so the equation becomes:

$$\frac{di}{dt} + 1000i = 50 \tag{2}$$

$$\Rightarrow \quad f(t, i) = \frac{di}{dt} = 50 - 1000i$$

Note that t does not appear in $f(t,i)$ though there is no reason why it should not do so in another problem. The initial condition is given as $i_A = 0.01$ when $t_A = 0$, so point A on the graph has the coordinates $(0, 0.1)$. Subscripts A, B ... refer to points A, B ... on the curve.

The gradient at point A, when $t_A = 0$, is:

$$\left.\frac{di}{dt}\right|_A = 50 - 1000 i_A = 50 - 1000 \times 0.01 = 40$$

The vertical line and A after di/dt symbolizes that we mean the particular value of di/dt at point A. If we place point B a little to the right of A, so that its t-coordinate is 0.00025, equivalent to $0.25\,\text{ms}$, we can calculate the value of its i-coordinate:

$$i_B = i_A + \left.\frac{di}{dt}\right|_A \cdot 0.0025 = 0.01 + 40 \times 0.00025$$

$$= 0.02$$

The basis of this calculation is shown in Figure 12.3, and point B is plotted in Figure 12.4 The interval 0.00025 is referred to as h in the box.

PRACTICAL DIFFERENTIALS 183

> **Euler's method**
>
> Given dy/dx = $f(x, y)$
>
> Plot point A (x_A, y_A) representing border condition
>
> Decide on suitable interval h between points, in x-direction.
>
> Calculate next point: $\quad B\left(x_A + h, \, y_A + h \cdot \left.\dfrac{dy}{dx}\right|_A\right)$
>
> Repeat to plot points in turn.
>
> Join the points.

We are now at point B, with coordinates (0.000 25, 0.02). The gradient here is:

$$\left.\frac{di}{dt}\right|_B = 50 - 1000 i_B = 50 - 1000 \times 0.02 = 30$$

$$\Rightarrow \quad i_C = i_B + \left.\frac{di}{dt}\right|_B \cdot 0.000\,25 = 0.02 + 30 \times 0.000\,25$$

$$= 0.0275$$

Point C is therefore (0.0005, 0.0275). Repeating this operation, we calculate points D, E, F, ... joining them by straight-line segments until we have plotted as much of the curve as required. Since the calculation repeats a short

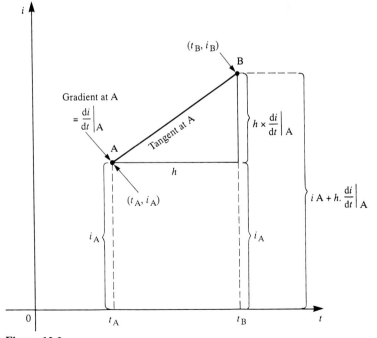

Figure 12.3

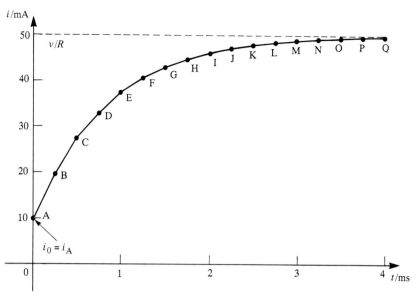

Figure 12.4

sequence of operations, a computer program only a few lines long readily provides the coordinates. Figure 12.4 shows the curve plotted using Euler's method. Compare this with the curve plotted in Figure 11.7. In this example, owing to the nature of the circuit, both curves approach the value v/R and therefore the difference between them becomes smaller and smaller as t increases. This may not happen with other equations, so this example gives a result more precise than is usual with Euler's method.

Test yourself 12.1

1 Given $\dfrac{dy}{dx} = y$, and the border condition, $y = 3$ when $x = 0$.

Using Euler's method, find y when $x = 4$, with $h = 1$. Repeat with $h = 0.05$ and $h = 0.25$. Compare with the result obtained by solving the equation exactly.

2 Given $\dfrac{dy}{dx} = 3x^2$, and the border condition, $y = 3$ when $x = 2$.

Using Euler's method, find y when $x = 2.28$, with $h = 0.4$. Compare with the result obtained by solving the equation exactly.

3 Given $\dfrac{dy}{dx} = x + y$, and the border condition, $y = 2$ when $x = 2$.

Using Euler's method, find y when $x = 3$, with $h = 0.2$. Confirm that the solution of this equation is $y = -x - 1 + 5e^{x-2}$, and use this solution to calculate the exact value of y for $x = 3$.

Operator D

Operator D is a symbol for the operation of differentiation:

- instead of writing $\dfrac{dy}{dx}$ we write Dy
- instead of writing $\dfrac{d^2y}{dx^2}$ we write D^2y
- in general, instead of writing $\dfrac{d^n y}{dx^n}$ we write $D^n x$

D can also be negative, which symbolizes the inverse operation, integration:

$$D^{-1}y \equiv \frac{1}{Dy} \equiv \int y\, dx$$

But operator D is more than just a short-hand way of saying 'differentiate'. It is the key to a powerful method of solving differential equations.

Properties of D

Given $D^n y$, we can differentiate it once more, to get $D^{n+1}y$, and again to get $D^{n+2}y$. We could differentiate it m times, to get $D^{n+m}y$. Differentiating $D^n y$ m times can also be written $D^m(D^n y)$. This shows that operator D obeys the index rule (page 24):

$$\mathbf{D^m(D^n y) = D^{m+n}y}$$

In this equation, D is behaving exactly as if it was an ordinary variable, such as we might find in the equation:

$$a^m(a^n y) = a^{m+n}y$$

A second property of the operator is the way it behaves with constants. When we differentiate a function such as ky, where k is a constant, the result is k times the differential of y:

$$\frac{d}{dx}(ky) = k \cdot \frac{dy}{dx}$$

In terms of operator D:

$$D(ky) = k \cdot D(y)$$

Again, D is behaving exactly as if it was an ordinary variable, such as we might find in the equation:

$$a(ky) = ka(y)$$

The variables a and k can be interchanged. So can D and k.

For operations such as differentiation and multiplication by a constant, D behaves like an ordinary variable. These properties are used to make it easier to process differential equations. For example, take the equation:

$$2\frac{d^2y}{dx^2} - \frac{dy}{dx} - 3y = 0$$

The left side of this is expressed in terms of operator D as:

$(2D^2 - D - 3)y$

If D behaves as an ordinary variable, it should be permissible to factorize the expression in brackets:

$(2D^2 - D - 3)y = (D + 1)(2D - 3)y$

Let us see if this works when the factorized expression is multiplied out again. We gradually replace the operator D symbols by their equivalents in ordinary differentiation symbols:

$(D + 1)(2D - 3)y = (D + 1)(2Dy - 3y)$

$= (D + 1)\left(2\dfrac{dy}{dx} - 3y\right)$

$= D\left(2\dfrac{dy}{dx} - 3y\right) + 1\left(2\dfrac{dy}{dx} - 3y\right)$

$= 2\dfrac{d^2y}{dx^2} - 3\dfrac{dy}{dx} + 2\dfrac{dy}{dx} - 3y$

$= 2\dfrac{d^2y}{dx^2} - \dfrac{dy}{dx} - 3y$

Now we are back to the original differential equation, and turn it into its operator D form:

$= 2D^2y - Dy - 3y$

$= (2D^2 - D - 3)y$

Multiplying out the factors leads us back to the original expression, showing that factorizing of expressions in D is a valid operation.

Using operator D

As an example, take a third-order equation:

$5\dfrac{d^3y}{dx^3} + 3\dfrac{d^2y}{dx^2} - 4\dfrac{dy}{dx} + 7y = 0$

The equation is clearer if we use operator D to express it. The example above may be written:

$(5D^3 + 3D^2 - 4D + 7)y = 0$

> **The meaning of $F(D)\,e^{ax}$**
>
> The examples below show how the equation works for some simple operations.
>
> *Using the equation to find the first derivative of e^{ax}:*
>
> By the definition of what we mean by D:
>
> $$\frac{d(e^{ax})}{dx} \equiv De^{ax}$$
>
> In this case the $F(D)$ we have to use is simply:
>
> D
>
> So the $F(a)$ we have to use is simply:
>
> a
>
> Now to use them:
>
> $$\frac{d(e^{ax})}{dx} \equiv De^{ax} = ae^{ax}$$
>
> Just replace D by a. The equation gives the same result as in the table of standard derivatives (page 122).
>
> *Similarly, we find the second derivative of e^{ax}, using the functions $F(D) = D^2$, then $F(a) = a^2$:*
>
> $$\frac{d^2(e^{ax})}{dx} \equiv D^2 e^{ax} = a^2 e^{ax}$$
>
> Again the equation gives the expected result. Always the function of a has exactly the same form as the function of D.
>
> *As a third example, we integrate e^{ax}, using the functions $F(D) = D^{-1}$, then $F(a) = a^{-1}$:*
>
> $$\int e^{ax}\,dx \equiv D^{-1} e^{ax} = a^{-1} e^{ax} = \frac{e^{ax}}{a}$$
>
> The result is the same as in the table of standard integrals, page 143.

The expression in brackets is a function of D, and the equation may be written even more compactly as:

$$f(D)y = 0$$

Thus $f(D)y$ may be thought of as the left side of any differential equation, complete with coefficients for each term. Some of the coefficients may be zero, in which cases there are missing terms.

One of the properties of functions such as the above, based on operator D, is expressed in the following identity:

$$F(D)e^{ax} \equiv e^{ax} F(a) \tag{1}$$

This gives us a way of differentiating expressions containing terms of the form e^{ax} (see box). Here is the proof of this identity:

| *The reasoning* | *The general term* |

$F(D)$ consists of a sequence of terms ... $a_n D^n$
(where a_n is the coefficient of the nth derivative)
So $F(D) e^{ax}$ is the sum of a series of terms ... $a_n D^n e^{ax}$
(a_n times the nth derivative of e^{ax})
Consider the e^{ax} part; differentiating once ... ae^{ax}
... differentiating again ... $a^2 e^{ax}$
... and differentiating n times ... $a^2 e^{ax}$
So $F(D)e^{ax}$ is the sum of the sequence ... $a_n a^n e^{ax}$
But e^{ax} is a constant factor in all these terms.
The sum is e^{ax} *times* the sequence ... $a_n a^n$
These terms have the same form as those of $F(D)$ $a_n D^n$
(see top line), except that they have a instead of D;
it is the *same function, F*. The sequence of terms
can be represented by $F(a)$.

It follows that:

Differentiating e^{ax} according to function $F(D)$ $\equiv$ **Multiplying e^{ax} by the function $F(a)$**

A related identity is an extension of the first:

$$F(D) \{e^{ax} f(x)\} \equiv e^{ax} F(D + a) f(x) \tag{2}$$

This may be proved by a similar argument. The box shows how it can be used for finding derivatives. The next example makes use of it for solving a differential equation.

The meaning of $F(D) \{e^{ax} f(x)\}$ ...

The simplest possible example has $F(D) = D$ and $f(x) = x$, and we try to find the first derivative of $e^{ax}x$:

$$\frac{d(e^{ax}x)}{dx} \equiv D\{e^{ax}x\}$$

By the equation, and using the function $F(D + a)$:

$$D\{e^{ax}x\} = e^{ax} \times (D + a) f(x)$$
$$= e^{ax} \times (D + a) x$$
$$= e^{ax} \times (Dx + ax)$$

But $Dx = 1$, so

$$D\{e^{ax}x\} = e^{ax} \times (1 + ax)$$
$$= e^{ax} + axe^{ax}$$

This result agrees with that obtained by differentiating $e^{ax}x$ using the Product Rule (page 130).

Solving a differential equation

On page 161 we introduced the decay equation modelling the charge on a capacitor. It was solved there by separating the variables. Now we will solve it by using operator D. The equation is:

$$\frac{dq}{dt} = \frac{-1}{RC} \cdot q$$

Rearrange the terms to bring them all to the left side:

$$\frac{dq}{dt} + \frac{1}{RC} \cdot q = 0$$

Writing Dq for dq/dt:

$$Dq + \frac{1}{RC} \cdot q = 0$$

Now treat this as an equation in which D is an ordinary variable. First take the common factor q outside the bracket:

$$\Rightarrow \quad \left(D + \frac{1}{RC} \right) q = 0$$

We need to simplify the expression in brackets. This is more easily done after multiplying throughout by $e^{t/RC}$ (see later for the reason):

$$\Rightarrow \quad e^{t/RC} \left(D + \frac{1}{RC} \right) q = 0 \quad (3)$$

Compare the left side of (3) item-by-item with the right side of (2) above:

In (3)	In (2)	Note
e	e	Same in both
t	x	Independent variable
$\frac{1}{RC}$	a	Constants as indices
	F	Not needed in (3), as the function is written out in full
D	D	Same in both
$\frac{1}{RC}$	a	Constants as terms
q	$f(x)$	Dependent vartiable

The left side of (3) corresponds exactly with the right side of (2).

In symbols:

$$e^{t/RC} \left(D + \frac{1}{RC} \right) q \quad \Leftrightarrow \quad e^{ax} F(D + a) x$$

left of (3) matches right of (2)

We know that right of (2) equals left of (2) and so:

$$F(D)\{e^{ax}f(x)\} \Leftrightarrow D(e^{t/RC}q)$$

left of (2)　　　matches　　new version of left of (3)

The new version of the left of (3) has the form of the left of (2):

$$\Rightarrow \quad D(e^{t/RC}q) = 0$$

This is simpler than the original version of (3), and therefore the subsequent calculations are made easier. The next step is to remove the D by integrating both sides, that is by multiplying both sides by D^{-1} (page 185):

$$\Rightarrow \quad e^{t/RC}q = A$$

Where A is the constant of integration.
Making q the subject:

$$\Rightarrow \quad q = Ae^{-t/RC}$$

This is the same result as obtained on page 161.

With such a short differential equation there is no advantage in using the operator D method. It is presented here as an application that is easy to understand. Solving this equation depends upon the apparent 'trick' of multiplying throughout by $e^{t/RC}$. This is the integrating factor described on page 164.

A related example

A slightly more complex equation appears on page 163:

$$\frac{di}{dt} + \frac{R}{L} \cdot i = \frac{v}{L}$$

This has a constant on the right instead of zero. This means that it cannot be solved by separating the variables. Using operator D, the calculation follows the same stages as that above. The integrating factor, is $e^{Rt/L}$ in this example. When multiplying by this factor, the constant on the right becomes

$$\frac{ve^{Rt/L}}{L}$$

After simplifying the left side of the equation, we have:

$$D(e^{Rt/L}i) = \frac{ve^{Rt/L}}{L}$$

In this example, when we remove D by integrating, we have to integrate the right side of the equation too:

$$e^{Rt/L}i = \frac{v}{L}\int e^{Rt/L}dt$$

$$= \frac{v}{R}e^{Rt/L} + A$$

where A is the constant of integration. Finally, dividing throughout by the integrating factor:

$$i = \frac{v}{R} + Ae^{Rt/L}$$

This is the same solution as was obtained before.

Solving second-order equations

On page 172 we solved the equation:

$$\frac{d^2y}{dx^2} - 3\frac{dy}{dx} + 2y = 0$$

Rewrite this in terms of operator D:

$$\Rightarrow \quad (D^2 - 3D + 2)y = 0$$

Factorize:

$$(D - 2)(D - 1)y = 0$$

There are two possible solutions to this. Either:

$$(D - 2)y = 0$$

or $\quad (D - 1)y = 0$

These equations may be taken separately and solved as above, with solutions:

$$y = Ae^{2x}$$
$$y = Be^x$$

The solution of the equation therefore includes both of these:

$$y = Ae^{2x} + Be^x$$

This is the same solution as obtained on page 172.

In this example, the function containing operator D could be factorized. This may not be possible in other examples but the roots of the function can be found by using the quadratic formula (page 39).

Equal roots

If the roots are equal, as in the example on page 173, the calculation goes as follows:

$$(D^2 + 2D + 1)y = 0$$

$$\Rightarrow \quad (D + 1)^2 y = 0$$

Multiply by the integrating factor and with a equal to 1, use equality (2), as on page 188:

$$e^x(D + 1)^2 y = 0$$

$$D^2(ye^x) = 0$$

To eliminate D^2 it is necessary to integrate twice. The first gives:

$$D(ye^x) = B$$

where B is the constant of the integration

The second gives:
$$ye^x = A + Bx$$

$$\underline{y = Ae^{-x} + Bxe^{-x}}$$

where A is the other constant of integration.

Imaginary roots

Consider this example:

$$\frac{d^2y}{dx^2} + 9y = 0$$

In terms of operator D:

$$(D^2 + 9)y = 0$$

To factorize this requires complex numbers (page 240), in which $j = \sqrt{-1}$

$$(D + j3)(D - j3) = 0$$

$$\Rightarrow \qquad y = Ae^{-j3x} + Be^{j3x}$$

Convert this from exponential form to polar form (page 249):

$$\Rightarrow \qquad y = A(\cos 3x - j \sin 3x) + B(\cos 3x + j \sin 3x)$$

$$= C \cos 3x + D \sin 3x$$

where $C = A + B$ and $D = j(A - B)$.

This result can be applied to any differential equation in which the operator consists of D^2 *plus* a square. If the number is not the square of an integer, we can use its square root:

$$\frac{d^2y}{dx^2} + 5y = 0$$

$$\Rightarrow \qquad \underline{y = C \cos(x\sqrt{5}) + D \sin(x\sqrt{5})}$$

Now look at this slightly extended example:

$$\frac{d^2y}{dx^2} - 4\frac{dy}{dx} + 29y = 0$$

In terms of operator D:

$$(D^2 - 4D + 29)y = 0$$

This is not D² plus a square. Factorizing, using a formula based on the quadratic formula $D + b/2a \pm j\left(\sqrt{|b^2 - 4ac|}\right)/2$:

$$(D - 2 + j5)(D - 2 - j5)y = 0$$

$\Rightarrow \qquad D = 2 - j5 \quad \text{or} \quad D = 2 + j5$

$\Rightarrow \qquad y = Ae^{(2 - j5)x} + Be^{(2 + j5)x}$

$\qquad\qquad = Ae^{2x} \cdot e^{-j5x} + Be^{2x} \cdot e^{j5x}$

$\qquad\qquad = e^{2x}(Ae^{-j5x} + Be^{j5x})$

Converting to polar form:

$$y = e^{2x}(C \cos 5x + D \sin 5x)$$

Higher-order equations

This is a third-order equation:

$$\frac{d^3y}{dx^3} - 4\frac{d^2y}{dx^2} - 17\frac{dy}{dx} + 60 = 0$$

Re-write using operator D:

$$(D^3 - 4D^2 - 17 + 60)y = 0$$

Factorize:

$$(D - 3)(D + 4)(D - 5) = 0$$

The three factors are treated just as the factors in previous examples. There are three solutions, each with its own constant of integration and the complete equation is:

$$y = Ae^{3x} + Be^{-4x} + Ce^{5x}$$

Now consider this equation:

$$\frac{d^5y}{dx^5} - \frac{d^4y}{dx^4} - \frac{dy}{dx} - y = 0$$

$\Rightarrow \qquad (D^5 - D^4 - D - 1)y = 0$

Factorizing:

$\Rightarrow \qquad (D - 1)^2 (D + 1)(D^2 + 1)y = 0$

There are several solutions, since the equation is satisfied when any one of a number of possible combinations of these factors equals zero.

- If $(D - 1)^2 y = 0$, we have equal roots (see above) giving the solution: $y = Ae^x + Bxe^x$
- If $(D + 1)y = 0$, the solution is Ce^{-x}
- If $(D^2 + 1)y = 0$, we have a 'D² plus a square' (see page 192) giving the solution: $y = D\cos x + E\sin x$.

Thus several of the different types of equation dealt with above are combined into one equation. For the complete solution, all these solutions are superposed:

$$y = Ae^x + Bxe^x + Ce^{-x} + D\cos x + E\sin x$$

This example begins to reveal the power of the operator D method for solving differential equations.

Test yourself 12.2

Use the operator D method to solve the following equations.

1. $\dfrac{dy}{dx} + 5y = 0$

2. $\dfrac{d^2y}{dx^2} + \dfrac{dy}{dx} - 6y = 0$

3. $\dfrac{d^2y}{dx^2} - 6\dfrac{dy}{dx} + 8y = 0$

4. $\dfrac{d^2y}{dx^2} + 4\dfrac{dy}{dx} + 4y = 0$

5. $\dfrac{d^2y}{dx^2} - 16y = 0$

6. $\dfrac{d^2y}{dx^2} + 16y = 0$

7. $\dfrac{d^2y}{dx^2} - 2\dfrac{dy}{dx} + 5y = 0$

8. $\dfrac{d^3y}{dx^3} + 4\dfrac{d^2y}{dx^2} - 7\dfrac{dy}{dx} + 10y = 0$

9. $\dfrac{d^4y}{dx^4} + 3\dfrac{d^2y}{dx^2} - 4y = 0$

10. $\dfrac{d^2y}{dx^2} + 4\dfrac{dy}{dx} + 13y = 0$

The Laplace transformation

The idea of making calculations easier by using a transformation is one we have met before. For example, suppose we wish to find the value of $4^{2.3}$. The way to do this is to take the log of 4, multiply this by 2.3 and then take the antilog of the product:

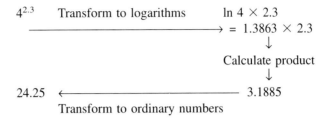

The operation begins with a transformation of ordinary numbers to logs and ends with a transformation of logs back into ordinary numbers. The reason for using this technique is that forming powers of numbers can be difficult, especially if the powers are fractional. But, after transforming the numbers into logs, the task is the much easier one of multiplication. The transformations from numbers to logs and from logs back to numbers are easily done using tables or a calculator.

Similarly, we can multiply two or more numbers together by taking their logs, summing them, and then taking the antilog of the sum. Before the days of pocket calculators this technique was commonly used to short-cut the process of long multiplication. It was also the basis of the slide rule.

We use the log transform to simplify the maths, so that we can multiply by adding, or form powers by multiplying. The reason for the Laplace transform is the same—to make calculations easier. Here the idea is to make it possible to *differentiate* a function by *multiplying* its transform.

The Laplace transform operates on a time function, usually represented by the symbol $f(t)$. The time function is usually an expression showing how some physical quantity, such as a current or a voltage, varies with time. The transform converts the function into a function of another variable s. The variable s is often a complex number (page 244) with a real part, Re(s) and an imaginary part Im(s).

The Laplace transform of $f(t)$ is symbolized by $\mathcal{L}[f(t)]$. It is also written as $F(s)$ and is defined by:

$$\mathcal{L}[f(t)] = F(s) = \int_0^\infty f(t)\, e^{-st}\, dt \tag{1}$$

In some books, the symbol p is used instead of s. Putting the equation above into words, we transform a time function $f(t)$ by integrating the product of this function and the function e^{-st} from zero to infinity. It may sound as if this could be a difficult thing to do but, as we shall explain later, it is made easy by using ready-made tables of transforms.

Certain conditions apply to this definition of the transform, the main one being that the integral must be **convergent**. That is to say, its value as t becomes large must approach some definite limiting value, rather than

> **Dimensions of s**
>
> The index of an exponential term must always be dimensionless. Thus the index $-st$ in the Laplace transformation must be dimensionless. The dimension of t is *time* (T). To make $-st$ dimensionless, the dimension of s must be the *reciprocal* of time ($1/T$). This is the dimension of *frequency*.
> The domain (page 72) of the untransformed function $f(t)$ is time. The domain of the transform $F(s)$ is frequency and, since s is very often a complex number, we say that s is in the **complex frequency domain**.

become increasingly large itself. The time functions most often met with in electronics are normally convergent when integrated, so this condition is not usually a problem. We also assume that $f(t) = 0$ when $t < 0$. This is generally true of circuits which have no currents flowing or voltages developed until the moment of switching on the power supply. If it is not true of a particular circuit, we introduce starting conditions into the calculation.

Transforming differential equations

In the discussion which follows, remember that all functions are time functions, symbol, $f(t)$ and all transforms are Laplace transforms, symbol, $F(s)$. Often we use symbols such as $i(t)$ and $v(t)$ to represent time functions of particular quantities such as current and voltage, and the symbols $I(s)$ and $V(s)$ to represent their transforms. Also, to make the calculations look clearer, we use the integral sign alone to mean the definite integral from zero to infinity:

$$\text{We write } \int \text{ to mean } \int_0^\infty$$

Strictly speaking, if the lower or upper limit (or both) of an integral is infinity, it can hardly be regarded as a *definite* integral. We call it an **improper integral**. However, it is usually possible to evaluate such an integral using the concept of limits (the other kind of limit, defined in Chapter 8).

Having analysed a circuit and modelled it as a differential equation, the next step is to transform the terms of that equation. Terms can be transformed individually because it can be shown that the transform of a sum of a number of functions equals the sum of the transforms of the individual functions. For example, a circuit is modelled by the second-order differential equation given on page 172.

$$\frac{d^2 i(t)}{dt^2} - 3 \frac{di(t)}{dt} + 2i(t) = 0$$

where the current varies in time according to the function $i(t)$. Using symbols appropriate to transforms, we have substituted $i(t)$ for y and t for x, but the equation is essentially the same. We have already solved this using the technique described on page 172, and also by using operator D (page 191). Now we will solve it by using the Laplace transform.

This equation has zero on the right, which is transformed as zero (see later for how to transform a non-zero constant). Begin by transforming the term in $i(t)$. We do not know yet what form this function takes, because we would have to solve the equation to find it. So we will just represent its transform by the symbol, $I(s)$. In the equation, the term in $i(t)$ has the coefficient 2. According to the rules of transformation, this coefficient also applies to the transform, so:

The transform of $2i(t)$ is $2I(s)$

To transform $di(t)/dt$, we operate according to equation (1):

$$\mathcal{L}\left[\frac{di(t)}{dt}\right] = \int \frac{di(t)}{dt} e^{-st}\, dt$$

This is integrated by parts (page 150):

$$u = e^{-st} \quad\Rightarrow\quad du/dt = -se^{-st}$$

$$dv/dt = di(t)/dt \quad\Rightarrow\quad v = i(t)$$

$$\int \frac{di(t)}{dt} e^{-st}\, dt = uv - \int v\, \frac{du}{dt} \cdot dt$$

$$= [e^{-st} i(t)]_0^\infty - \int_0^\infty i(t) - se^{-st}\, dt$$

$$= -i(0) + s\int_0^\infty i(t)\, e^{-st}\, dt$$

When evaluated as a definite integral, $e^{-st}i(t)$ becomes $-i(t)$. This is its value when $t = 0$ so we represent it by $-i(0)$, the current flowing the *instant* after timing begins.

The last expression above is the same as $i(t)$ transformed into $I(s)$ by applying equation (1), and then multiplying by s. In other words, the transform of $di(t)/dt$ is $sI(s) - i(0)$ and, taking the coefficient into account, we find that the transform of

$$3\, \frac{di(t)}{dt}$$

is $3[sI(s) - i(0)]$. Note that, instead of *differentiating* $i(t)$, we merely have to *multiply* its transform by s. In this example, we also specify the border conditions, which are those obtained when $t = 0$. If there is 1 A flowing at that time, then $i(0) = 1$. This simplifies the transform:

The transform of $-3\,\dfrac{di(t)}{dt}$ is $-3sI(s) + 3$

Transforming differential equations

Term	Transformed term
Constant a	a/s
Function $f(t)$	$F(s)$
1st derivative	$sF(s) - f(0)$
2nd derivative	$s^2 F(s) - sf(0) - \dfrac{df(0)}{dt}$

The transform of a second derivative $d^2 i(t)/dt^2$ is an extension of the transformation of a first derivative:

$$\mathcal{L}[d^2 i(t)/dt] = s\{sI(s) - i(0)\} - \frac{di(0)}{dt}$$

$$= s^2 I(s) - si(0) - \frac{di(0)}{dt}$$

The last expression is the value of the *differential* or rate of change of i when $t = 0$. Suppose that this is zero, as it probably would be in a circuit at the instant power is switched on.

Substituting $\dfrac{di(0)}{dt} = 0$ and $i(0) = 1$:

The transform of $\dfrac{d^2 i(t)}{dt^2}$ is $s^2 I(s) - s$

Now we are ready to sum the transforms to obtain the transformed equation:

$$s^2 I(s) - s - 3sI(s) + 3 + 2I(s) = 0$$

The transformation is complete. The next stage is to work on it, simplifying it as far as possible. Rearranging and gathering terms:

$$(s^2 - 3s + 2) I(s) = s - 3$$

Note that the coefficients in the brackets of the first term are the same as in the original equation. Making $I(s)$ the subject:

$$I(s) = \frac{s - 3}{(s^2 - 3s + 2)}$$

Factorizing the denominator:

$$I(s) = \frac{s - 3}{(s - 1)(s - 2)}$$

In most calculations of this kind the next step is to replace the expression on the right by the equivalent partial fractions. The technique for this is explained on page 18 and the result is:

$$I(s) = \frac{2}{s-1} - \frac{1}{s-2}$$

The reason for putting the equation into partial fractions is that it makes it easier to transform the equation back into ordinary numbers again. This inverse transformation is the final step of the calculation. We leave this example waiting while we discuss the inverse transformation technique.

Inverse transformation

The algebra of inverse transformation is very involved but, fortunately, there is a way of avoiding it. This relies on the property of the Laplace transform that the functions and their transforms exist in one-to-one pairs. Going from function to transform, there is only one transform. Similarly at the end of a calculation, when we go from transform back to function, there is only one way the reverse transformation can be made. The approach to the problem of inverse transformation is to construct a table of function–transform pairs. This is relatively easy, using equation (1). In this way we build up a table, listing common functions and their transforms. Then, when the time comes to perform the inverse operation, we look up the transform in the table, which then takes us back to the corresponding function.

Below we look at several common functions and find their transforms. As before, all the integrals are improper integrals summed between the limits $t = 0$ to $t = \infty$.

The unit step function

$$\mathcal{L}[u(t)] = \int 1 e^{-st} \, dt = \frac{-1}{s} [e^{-st}]_0^\infty = \frac{1}{s}$$

This follows the usual method for evaluation of a definite integral. When t approaches infinity, e^{-st} approaches zero; when $t = 0$, then $e^{-st} = e^0 = 1$; the expression in square brackets on the right equals -1.

Since the transform of a function multiplied by a constant is the transform multiplied by that constant, the transform of a non-unit step function such as $au(t)$ is:

$$\mathcal{L}[au(t)] = \frac{a}{s}$$

A constant term in a differential equation is equivalent to a step. This is why the transform of a constant a is a/s.

Step function

There is an instantaneous increase of y

The **unit step function** $u(t)$ is an increase of 1, usually when time $t = 0$.

The function is defined as:

$u(t) = 0 \qquad -\infty < t < 0^-$

$u(t) = 1 \qquad 0^+ < t\, \infty$

$t = 0^-$ at the instant before the change occurs
$t = 0^+$ at the instant after the change occurs
$u(t)$ is undefined at $t = 0$

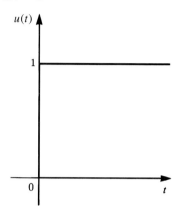

Figure 12.5

A step up to some other value, such as V volts is $Vu(t)$.
The function can have a negative value (a step down).

The ramp function

$$\mathscr{L}[t] = \int_0^\infty t e^{-st}\, dt \tag{2}$$

This is integrated by parts:

Let $u = t \quad\Rightarrow\quad du/dt = 1$

Let $dv/dt = e^{-st} \quad\Rightarrow\quad v = \dfrac{-1}{s} e^{-st}$

$$\Rightarrow \quad \int t e^{-st}\, dt = uv - \int v \cdot \frac{du}{dt}\, dt$$

$$= \frac{-t}{s} e^{-st} - \int \frac{-1}{s} e^{-st} \cdot 1\, dt$$

> **Ramp function**
>
> A steady increase with time
>
> The function is defined as:
>
> $f(t) = t$
>
> The function can be negative (a ramp down)
>
>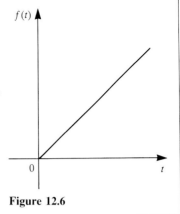
>
> Figure 12.6

Substituting this result in equation (2):

$$\mathcal{L}[t] = \left[\frac{-t}{s}e^{-st}\right]_0^\infty - \int_0^\infty \frac{-1}{s}e^{-st} \cdot 1 \, dt$$

$$= 0 + \frac{1}{s}\left[\frac{1}{s}e^{-st}\right]_0^\infty = \frac{1}{s^2}$$

The exponential growth function

If a is constant:

$$\mathcal{L}[e^{at}] = \int_0^\infty e^{at} e^{-st} \, dt = \int_0^\infty e^{(a-s)t} \, dt = \int_0^\infty e^{-(s-a)t} \, dt$$

$$= \frac{-1}{s-a}\left[e^{-(s-a)t}\right]_0^\infty = \frac{1}{s-a}$$

> **Exponential function**
>
> An exponential increase (growth) with time (see Figure 5.15, page 59). Or an exponential decrease (decay).
>
> The function is defined as:
>
> $f(t) = e^{at}$
>
> where a is a constant. The function represents decay if the index is negative.

The integral converges only if the real part of s is greater than the real part of a. If it is less, then the index of $e^{-(s-a)t}$ is positive and increases with time, making $e^{-(s-a)t}$ increase with time.

We transform the exponential decay function in a similar way, obtaining:

$$\mathcal{L}[e^{-at}] = \frac{1}{s+a}$$

The sine function

In the expression, ω is the angular velocity (page 95).

$$\mathcal{L}[\sin \omega t] = \int_0^\infty (\sin \omega t)\, e^{-st}\, dt \tag{3}$$

Before integrating this, consider the following expression:

$$e^{j\omega t} - e^{-j\omega t}$$

in which $j = \sqrt{-1}$. This may be written in polar form (page 249):

$$(\cos \omega t + j \sin \omega t) - (\cos \omega t - j \sin \omega t)$$

$$= \cos \omega t + j \sin \omega t - \cos \omega t + j \sin \omega t$$

$$= 2j \sin \omega t$$

$$\Rightarrow \quad \sin \omega t = \frac{1}{j2}(e^{j\omega t} - e^{-j\omega t})$$

The sine function is $1/j2$ times the difference between two exponential functions. The transform of the sine function is thus $1/j2$ times the difference between the transforms of these functions:

$$\mathcal{L}[\sin \omega t] = \frac{1}{j2}\left\{\mathcal{L}[e^{j\omega t}] - \mathcal{L}[e^{-j\omega t}]\right\}$$

Using the transforms for growth and decay which we have already found:

$$= \frac{1}{j2}\left\{\frac{1}{s-j\omega} - \frac{1}{s+j\omega}\right\}$$

$$= \frac{1}{j2}\left\{\frac{s+j\omega - s + j\omega}{s^2 + \omega^2}\right\}$$

$$= \frac{1}{j2} \cdot \frac{j2\omega}{s^2 + \omega^2}$$

$$= \frac{\omega}{s^2 + \omega^2}$$

Note how, having obtained a few easy transforms, we use these to find other more difficult transforms.

–S12.9–

Laplace transform pairs

a, Φ and ω are constants

$f(t)$	$F(s)$	Conditions
1 (unit step function)	$1/s$	$Re(s) > 0$
a (step function)	a/s	$Re(s) > 0$
t (ramp function)	$1/s^2$	
e^{at} (growth function)	$\dfrac{1}{s-a}$	$Re(s) > a$
e^{-at} (decay function)	$\dfrac{1}{s+a}$	$Re(s) > -a$
$1 - e^{at}$	$\dfrac{a}{s(s-a)}$	$Re(s) > a$
$1 - e^{-at}$	$\dfrac{a}{s(s+a)}$	$Re(s) > -a$
$\sin \omega t$	$\dfrac{\omega}{s^2 + \omega^2}$	$s > 0$
$\cos \omega t$	$\dfrac{s}{s^2 + \omega^2}$	$s > 0$
$\sin(\omega t + \Phi)$	$\dfrac{s \sin \Phi + \omega \cos \Phi}{s^2 + \omega^2}$	$s > 0$
$e^{-at} \sin \omega t$	$\dfrac{\omega}{(s+a)^2 + \omega^2}$	
$e^{-at} \cos \omega t$	$\dfrac{s+a}{(s+a)^2 + \omega^2}$	

Continuing the example

On page 199 we left the example as a transform consisting of two partial fractions:

$$\frac{2}{s-1} \quad \text{and} \quad \frac{-1}{s-2}$$

Scanning the second column of the table, we find that these fractions have the form:

$$\frac{1}{s-a}$$

which is the transform of the growth function e^{at}. In the first fraction, $a = 1$, and the transform is multiplied by a constant 2. In the second fraction, $a = 2$, and the constant multiplier is -1. Making the inverse transformation, we arrive at the solution of the differential equation:

$$\underline{i(t) = 2e^t - e^{2t}}$$

On page 172 the solution is given as $y = Ae^{2x} + Be^x$. However, if we use the border conditions given above, we find that $A = -1$ and $B = 2$, so we have obtained an identical result.

> **Solving differential equations using Laplace transforms – a summary**
>
> 1 Transform the terms of the equation.
> 2 Operate on the transformed equation to simplify it, and insert initial values.
> 3 Try to obtain expressions that match the standard transforms. This may often mean forming partial equations.
> 4 Inverse-transform the equation to obtain the solution.

Another example

Solve this second-order differential equation:

$$2\frac{d^2x}{dt^2} + 7\frac{dx}{dt} + 6x = 4$$

Transforming according to the rules (page 198):

- The constant, being four times the unit step function, becomes $4/s$.
- The third term, with coefficient 6, becomes $6F(s)$.
- The first derivative becomes $7[sF(s) - f(0)]$.
- The second derivative with coefficient 2, becomes

$$2\left[s^2 F(s) - sf(0) - \frac{df(0)}{dt}\right]$$

The complete transformed equation, after collecting terms, is:

$$[2s^2 + 7s + 6]F(s) = 4/s + (2s + 7)f(0) + \frac{7df(0)}{dt}$$

If border conditions for $t = 0$ are known, the equation can be simplified further by substituting for $f(0)$ and $df(0)/dt$. For example, suppose that both of these are zero, modelling a circuit with no current flowing at the instant when the power is applied. The equation becomes:

$$[2s^2 + 7s + 6]F(s) = 4/s$$

$$\Rightarrow \quad F(s) = \frac{4}{s(2s^2 + 7s + 6)}$$

Factorizing:

$$= \frac{4}{s(2s + 3)(s + 2)}$$

Forming partial fractions:

$$= \frac{2}{3s} - \frac{16}{3(2s+3)} + \frac{2}{s+2}$$

$$= \frac{2}{3}\left[\frac{1}{s} - \frac{4}{s+1.5} + \frac{3}{s+2}\right]$$

Ignoring constants, the fraction have forms that can be found in the table of transform pairs:

- 1st fraction is of the form $1/s \Rightarrow$ unit step function
- 2nd fraction is of the form $1/(s-a) \Rightarrow$ growth function
- 3rd fraction is of the form $1/(s+a) \Rightarrow$ decay function.

The inverse transform gives:

$$i(t) = \frac{2}{3}[1 - 4e^{1.5t} + 3e^{-2t}]$$

This is the solution to the differential equation.

More transforming routines

This example illustrates some further techniques used in solving a second-order differential equation:

$$\frac{d^2 i}{dt^2} + 6\frac{di}{dt} + 25i = 100$$

We are given that both initial conditions are zero. Writing out the transform gives:

$$s^2 I(s) - sI(0) - \frac{di(0)}{dt} + 6sI(s) - 6I(0) + 25I(s) = \frac{100}{s}$$

But, since initial conditions are zero, this simplifies to:

$$s^2 I(s) + 6sI(s) + 25I(s) = \frac{100}{s}$$

$$\Rightarrow \quad I(s) = \frac{100}{s(s^2 + 6s + 25)}$$

In this example, the term in brackets does not factorize. Since the term contains s^2, the partial fractions (page 18) are of the form:

$$\frac{A}{s} + \frac{Bs + C}{s^2 + 6s + 25}$$

Multiplying out:

$$100 = A(s^2 + 6s + 25) + s(Bs + c)$$

$$= As^2 + 6As + 25A + Bs^2 + Cs$$

Equating coefficients:

$$A + B = 0 \qquad 6A + C = 0 \qquad 25A = 100$$

From which we obtain:

$$A = 4 \qquad B = -4 \qquad C = -24$$

Substituting these values:

$$I(s) = \frac{4}{s} + \frac{-4s - 24}{s^2 + 6s + 25}$$

Before this can be inverse-transformed it must be moulded into one or more of the standard forms. There is no difficulty with the first term as this has the form of the step function and becomes 4.

The denominator of the second term does not factorize readily. Since many of the transforms on page 203 contain s^2 or $(s + a)^2$ in the denominator, we might try to find a square in $(s^2 + 6s + 25)$. The coefficient of s gives a clue: try $(s + 3)^2$. This when squared yields $(s^2 + 6s + 9)$ so the denominator can be broken into $(s + 3)^2 + 16$. The 16 is itself a square so (assuming that the problem is set so as to provide integers in the result) we have an expression of the same form as $(s + a)^2 + \omega^2$. The transforms which have this in the denominator are derived from $e^{-at} \sin \omega t$ and $e^{-at} \cos \omega t$.

Manipulate this numerator so that it contains $(s + 3)$:

$$-4s - 24 = -4(s + 3) - 12$$

This means that the second fraction is to be broken into two, and the complete expression becomes:

$$I(s) = \frac{4}{s} - \frac{4(s + 3)}{(s + 3)^2 + 16} - \frac{12}{(s + 3)^2 + 16}$$

The equation is ready for reverse transformation:

- $4/s$ becomes 4;
- in the second term, there is a constant multiplier ($= 4$), and the term is the transform of $e^{-at} \cos \omega t$, with $a = 3$ and $\omega = 4$;
- in the third term, there is a constant multiplier ($=3$), and the term is the transform of $e^{-at} \sin \omega t$, with $a = 3$ and $\omega = 4$.

Thus, after inverse transformation, the solution to the equation is:

$$\underline{i(t) = 4 - 4e^{-3t} \cos 4t - 3e^{-3t} \sin 4t}$$

Transforming integrals

Sometimes an equation may contain one or more integrals. This applies particularly when the modelled circuit includes capacitors, for the charge on a capacitor at time t is given by the expression:

$$q(T) = \int_0^t i(T) \, dT + q(0)$$

PRACTICAL DIFFERENTIALS

$i(T)$ is the function describing how the charging current varies with time; we use the symbol T as the time variable in this equation so as to distinguish it from t the instant of time which is the upper limit of the integral. $q(0)$ is the charge, if any, on the capacitor at the start, when $t = 0$.

The transform of an integral can be found in the usual way, by applying equation (1). Without going into details, the transform of the integral of a function is:

$$\mathcal{L}\left[\int_0^x f(x)\,\mathrm{d}x\right] = \frac{F(s)}{s} + \frac{f(0)}{s}$$

On page 197 we found that differentiating a function is equivalent to multiplying its transform by s. Here we see that integrating a function is equivalent to *dividing* its transform by s.

Example

A 50 µF capacitor is wired in series with a 25 Ω resistor. A voltage of 50 V is connected to the circuit when $t = 0$. The initial charge on the capacitor is $q(0) = 10$ mC, with polarity *opposite* to that of the charging voltage $i(0) = 0$. By applying KVL we obtain the equation modelling the flow of current into the capacitor:

$$Ri(T) + \frac{1}{C}\left[\int_0^t i(T)\,\mathrm{d}T + q(0)\right] = V$$

Substituting given values:

$$25i(T) + \frac{1}{50 \times 10^{-6}}\left[\int_0^t i(T)\,\mathrm{d}T + (-10 \times 10^{-3})\right] = 50$$

$$\Rightarrow \quad 25i(T) + \frac{-10 \times 10^{-3}}{50 \times 10^{-6}} + \frac{1}{50 \times 10^{-6}}\int_{-0}^t i(T)\,\mathrm{d}T = 50$$

$$\Rightarrow \quad 25i(T) - 200 + \frac{1}{50 \times 10^{-6}}\int_0^t i(T)\,\mathrm{d}T = 50$$

$$\Rightarrow \quad 25i(T) + \frac{1}{50 \times 10^{-6}}\int_0^t i(T)\,\mathrm{d}T = 250$$

The transform of $25i(t)$ is $25I(s)$

The transform of $\dfrac{1}{50 \times 10^{-6}}\displaystyle\int_0^t i(T)\,\mathrm{d}T$ is $\dfrac{I(s)}{50s \times 10^{-6}}$

Omitting $f(0)/s$, as $i(0) = 0$.

Note the s in the divisor. The transform of the constant 250 is $\dfrac{250}{s}$

Assembling the above, the transform of the equation is:

$$25I(s) + \frac{I(s)}{50s \times 10^{-6}} = \frac{250}{s}$$

Solving the transformed equation:

$$\Rightarrow \quad I(s)\left(25 + \frac{1}{50s \times 10^{-6}}\right) = \frac{250}{s}$$

$$\Rightarrow \quad I(s)\left(\frac{0.00125s + 1}{50s \times 10^{-6}}\right) = \frac{250}{s}$$

$$\Rightarrow \quad I(s) = \frac{250}{s} \times \frac{50s \times 10^{-6}}{0.00125s + 1}$$

$$\Rightarrow \quad I(s) = \frac{0.0125}{0.00125s + 1}$$

Multiply numerator and denominator by 800, to obtain a quotient suitable for inverse-transforming:

$$\Rightarrow \quad I(s) = \frac{10}{s + 800}$$

The inverse transform of this (page 203) is:

$$\underline{i(t) = 10e^{-800t}}$$

This is the solution to the equation.

Initial value

Although we are often *given* the initial value $f(0)$ as one of the boundary conditions, there are occasions when we need to be able to calculate it. We use the **initial value theorem**:

$$f(0) = \lim_{s \to \infty} sF(s)$$

Take as an example a circuit for which:

$$v = 3e^{-t} \qquad (4)$$

With such an elementary equation the initial voltage can be calculated directly by substituting $t = 0$:

$$v(0) = 3e^0 = 3$$

This problem can also be solved using the initial value theorem. First we transform the function of equation (4), using the transform for the decay function, with $a = 1$:

$$V(s) = \frac{3}{s + 1}$$

Apply the initial value theorem, using the method on page 107:

$$v(0) = \lim_{s \to \infty} sV(s) = \lim_{s \to \infty} s\left[\frac{3}{s + 1}\right] = 3$$

This is the same result as was obtained directly.

As a slightly more complicated example, suppose that the equation modelling the current in a circuit has been transformed and gives:

$$I(s) = \frac{2s + 1}{s + 2}$$

Applying the theorem:

$$i(0) = \lim_{s \to \infty} sI(s) = \lim_{s \to \infty} s \left[\frac{2s + 1}{s - 2} \right] = 2$$

The initial current is 2 A.

Final value

Corresponding to the theorem described above, there is also the **final value theorem**. This tells us the final (steady-state) value reached after a period of transition:

$$f(\infty) = \lim_{s \to 0} sF(s)$$

For example, the transform given in the second example above is:

$$I(s) = \frac{2s + 1}{s - 2}$$

Applying the theorem:

$$i(\infty) = \lim_{s \to 0} sI(s) = \lim_{s \to 0} s \left[\frac{2s + 1}{s - 2} \right] = \frac{1}{-2} = -0.5$$

The final current is –0.5 A.

Consider this example, for a circuit in which the transform of a voltage is:

$$V(s) = \frac{6}{s(s + 4)}$$

Applying the theorem:

$$v(\infty) = \lim_{s \to 0} sV(s) = \lim_{s \to 0} s \left[\frac{6}{s(s + 4)} \right]$$

$$= \lim_{s \to 0} \frac{6}{s + 4}$$

$$= \frac{6}{4} = 1.5$$

The final voltage is 1.5 V.

Test yourself 12.3

Calculate the Laplace transform of the following functions, using the defining equation (1) on page 195.

1 5 **2** $3t$ **3** e^{-2t} **4** $\cos 3t$

Using the table on page 203, find the inverse Laplace transform of the following functions.

5 $\dfrac{4}{s^2}$ **6** $\dfrac{3}{s+7}$ **7** $\dfrac{s+2}{(s+2)^2+25}$

8 $\dfrac{s+3}{s^2+6s+25}$ **9** $\dfrac{3}{s(s+3)}$ **10** $\dfrac{s \sin 2 + 3 \cos 2}{s^2+9}$

11 $\dfrac{s+2}{s^2+4}$ **12** $\dfrac{5s-4}{s^2-s-2}$ **13** $\dfrac{5s+20-2s^2}{s^3+4s^2}$

Write the Laplace transform of the following differential equations. Solve each equation, using the starting conditions given.

14 $\dfrac{dv}{dt} + 5v = 0$ **15** $\dfrac{d^2i}{dt^2} + \dfrac{di}{dt} - 6i = 0$

given that $v(0) = 2$ given that $i(0) = 0$ and $\dfrac{di(0)}{dt} = 5$

16 $2\dfrac{di}{dt} + 4i = 0$ **17** $2i + \dfrac{1}{2}\int_0^t i(T)\,dT = 4$

given that $i(0) = 2$ given that $i(0) = 0$

18 In the circuit of Figure 12.7a, the initial voltage across the capacitor is zero. A steady current of 2 A is supplied from $t = 0$.

  a Express the voltage across this circuit (i.e. at point A) as a differential equation. [Hint: use KCL at point A.]

  b Find the Laplace transform of the voltage equation and simplify it.

  c Make the inverse transform to find out how voltage varies with time.

19 In the circuit of Figure 12.7b, the initial applied voltage is zero, increasing by 2 V per second. The initial current is zero.

  a Express the current in this circuit as a differential equation.

  b Find the Laplace transform of the current equation and simplify it.

  c Make the inverse transform to find how current varies with time.

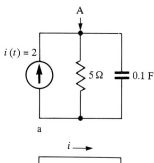

a

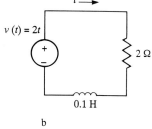

b

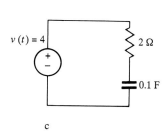

c

Figure 12.7

20 In the circuit of Figure 12.7c, the initial voltage across the capacitor is zero. A steady voltage is applied from time $t = 0$.

  a Express the current in this circuit as a differential equation.

  b Find the Laplace transform of the current equation and simplify it.

  c Make the inverse transform to find how the current varies with time.

21 Use the initial value theorem to find the initial and final voltages or currents in the circuits of Figure 12.7.

22 The current in a circuit is represented in the *s*-domain by:
$$I(s) = \frac{5}{2s + 3}$$
Calculate the initial and final currents.

23 The current in a circuit is represented in the *s*-domain by:
$$I(s) = \frac{9}{(s + 2)(s + 3)}$$
Calculate the initial and final currents.

13 Systematic solutions

> Simultaneous equations are used a lot in electronics, especially for analysing currents in networks. Here we look at determinant and matrix arithmetic, which provide the basis of two routines for solving such equations.
>
> You need to know about indices (page 23), balancing equations (page 34), changing the subject (page 37) and simultaneous equations (page 41).

Two familiar ways of solving simultaneous equations are by eliminating coefficients and by substitution. Both methods are useful for equations with two or three variables, but have their disadvantages. The elimination method requires a knack for spotting ways of getting rid of the coefficients with the minimum of calculating. The substitution method can lead to lengthy calculations and awkward fractions, and hence the increased possibility of making mistakes. Both methods become very unwieldy when the results are not integers or there are more than three variables. In this chapter we look at two systematic techniques for dealing with the more difficult sets of simultaneous equations. We shall illustrate them by solving simple examples with only two or three variables, remembering that these techniques really demonstrate their power only with more difficult equations.

Elimination in theory

Suppose we have a pair of equations, with two variables x and y:

$$a_1 x + b_1 y = c_1 \qquad (1)$$

$$a_2 x + b_2 y = c_2 \qquad (2)$$

Instead of numeric coefficients we have used constants a_1, b_1 and so on, so the results of the discussion which follows apply to *all* pairs of simultaneous equations. To solve these by the elimination method, eliminating y:

Equation (1) $\times b_2 \Rightarrow$

$$a_1 b_2 x + b_1 b_2 y = c_1 b_2 \qquad (3)$$

Equation (2) $\times b_1 \Rightarrow$

$$a_2 b_1 x + b_1 b_2 y = c_2 b_1 \qquad (4)$$

Subtract (4) from (3):

$$(a_1 b_2 - a_2 b_1) x = c_1 b_2 - c_2 b_1$$

$$x = \frac{c_1 b_2 - c_2 b_1}{a_1 b_2 - a_2 b_1} \qquad (5)$$

This solution holds true provided that the denominator is not equal to zero, for this would give an indeterminate result. Similarly, to eliminate x, we multiply (1) by a_2 and (2) by a_1:

$$a_1 a_2 x + a_2 b_1 y = a_2 c_1 \tag{6}$$

$$a_1 a_2 x + a_1 b_2 y = a_1 c_2 \tag{7}$$

Subtract (6) from (7):

$$(a_1 b_2 - a_2 b_1) y = a_1 c_2 - a_2 c_1$$

$$y = \frac{a_1 c_2 - a_2 c_1}{a_1 b_2 - a_2 b_1} \tag{8}$$

Again, the denominator must not equal zero.

The solutions for x and y both have exactly the same denominator, and their numerators have similar form. In fact, numerators and denominators all have the same form to the extent that the subscripts are 1 and 2 in the first term, and are 2 and 1 in the second term. We can set out the coefficients of the denominator in a table, like this:

$$\begin{matrix} a_1 & b_1 \\ a_2 & b_2 \end{matrix}$$

The denominator is calculated by taking the product of the terms on the leading diagonal ($a_1 b_2$) and subtracting the product of the terms on the other diagonal ($a_2 b_1$). To indicate that this is what has to be done with the terms in the table, we draw lines down each side of the table:

$$\begin{vmatrix} a_1 & b_1 \\ a_2 & b_2 \end{vmatrix}$$

A table of coefficients or constants set out like this is called a **determinant**. The coefficients are called the **elements** of the determinant. The number of rows in a determinant is always equal to the number of columns. The example above has two rows and two columns, so it is known as a **second-order determinant**.

Evaluating determinants

Here are some examples of determinants and their values:

(a) $\begin{vmatrix} 2 & 4 \\ 1 & 3 \end{vmatrix}$

The product of the elements on the leading diagonal is $2 \times 3 = 6$.
The product of the elements on the other diagonal is $1 \times 4 = 4$.
The value of the determinant is $6 - 4 = 2$.

(b) $\begin{vmatrix} 2 & 3 \\ 5 & 4 \end{vmatrix}$

The product on the leading diagonal is $2 \times 4 = 8$.
The product on the other diagonal is $5 \times 3 = 15$.
The value of the determinant is $8 - 15 = -7$.

(c) $\begin{vmatrix} 3 & 1 \\ 7 & -2 \end{vmatrix}$

The product on the leading diagonal is $3 \times -2 = -6$.
The product on the other diagonal is $7 \times 1 = 7$.
The value of the determinant is $-6 - 7 = -13$.

> **Test yourself 13.1**
>
> Find the values of these determinants.
>
> **1** $\begin{vmatrix} 1 & 3 \\ 2 & 7 \end{vmatrix}$ **2** $\begin{vmatrix} 3 & 4 \\ 2 & 5 \end{vmatrix}$ **3** $\begin{vmatrix} 5 & 3 \\ 7 & 2 \end{vmatrix}$
>
> **4** $\begin{vmatrix} 2 & 3 \\ -5 & 2 \end{vmatrix}$ **5** $\begin{vmatrix} -3 & 7 \\ 4 & 6 \end{vmatrix}$ **6** $\begin{vmatrix} 8 & -3 \\ -3 & 4 \end{vmatrix}$

Solving equations by using determinants

Now that we have defined the meaning of the determinant format, we can rewrite the solutions of a pair of simultaneous equations in determinant form:

$$x = \frac{\begin{vmatrix} c_1 & b_1 \\ c_2 & b_2 \end{vmatrix}}{\begin{vmatrix} a_1 & b_1 \\ a_2 & b_2 \end{vmatrix}} \qquad y = \frac{\begin{vmatrix} a_1 & c_1 \\ a_2 & c_2 \end{vmatrix}}{\begin{vmatrix} a_1 & b_1 \\ a_2 & b_2 \end{vmatrix}}$$

As before, the denominators must not have a zero value. We will use these equations to solve the equations previously given on page 41:

$$2x + y = 9$$
$$9x - y = 13$$

Comparison with equations (1) and (2) above gives the following equalities:

$a_1 = 2$ $b_1 = 1$ $c_1 = 9$
$a_2 = 9$ $b_2 = -1$ $c_2 = 13$

Written using determinants, the solutions are:

$$x = \frac{\begin{vmatrix} 9 & 1 \\ 13 & -1 \end{vmatrix}}{\begin{vmatrix} 2 & 1 \\ 9 & -1 \end{vmatrix}} \qquad y = \frac{\begin{vmatrix} 2 & 9 \\ 9 & 13 \end{vmatrix}}{\begin{vmatrix} 2 & 1 \\ 9 & -1 \end{vmatrix}}$$

Evaluating the determinants:

$$x = \frac{-22}{-11} \qquad y = \frac{-55}{-11}$$

Simplifying the fractions:

$$\underline{x = 2 \quad \text{and} \quad y = 5.}$$

These are the same results as were obtained by elimination.

Although we have used symbols such as a_1, b_2, and others for the elements, so as to follow the mechanism of evaluating determinants and solving equations, the method can be reduced to a simple rule of thumb, as shown in the box.

Solving simultaneous equations with two variables

$$x = \frac{\begin{bmatrix} \text{Values} \\ \text{of} \\ \text{constants} \end{bmatrix} \begin{bmatrix} \text{Coefficients} \\ \text{of} \\ y \end{bmatrix}}{\begin{bmatrix} \text{Coefficients} \\ \text{of} \\ x \end{bmatrix} \begin{bmatrix} \text{Coefficients} \\ \text{of} \\ y \end{bmatrix}}$$

$$y = \frac{\begin{bmatrix} \text{Coefficients} \\ \text{of} \\ x \end{bmatrix} \begin{bmatrix} \text{Values} \\ \text{of} \\ \text{constants} \end{bmatrix}}{\begin{bmatrix} \text{Coefficients} \\ \text{of} \\ x \end{bmatrix} \begin{bmatrix} \text{Coefficients} \\ \text{of} \\ y \end{bmatrix}}$$

Applying the rules to the equations from page 42:

$2x + y = 3$

$x - 3y = 19$

Coefficients of x are 2
  1

Coefficients of y are 1
  -3

Constants are 3
  19

Inserting these in the formats shown in the box:

$$x = \frac{\begin{vmatrix} 3 & 1 \\ 19 & -3 \end{vmatrix}}{\begin{vmatrix} 2 & 1 \\ 1 & -3 \end{vmatrix}} \qquad y = \frac{\begin{vmatrix} 2 & 3 \\ 1 & 19 \end{vmatrix}}{\begin{vmatrix} 2 & 1 \\ 1 & -3 \end{vmatrix}}$$

$$x = \frac{-28}{-7} \qquad y = \frac{35}{-7}$$

$$\underline{x = 4 \quad \text{and} \quad y = -5.}$$

Test yourself 13.2

Use determinants to solve these pairs of simultaneous equations:

1. $3x + 4y = 17$
   $4x - 3y = 16$

2. $2x + y = 7$
   $3x + 2y = 10$

3. $2x - 3y = 11$
   $x + y = -7$

4. $3x - 7y = 0$
   $2x - 3y = 5$

5. $2x - 5y = 3$
   $3x + 10y = 8$

6. In the circuit of Figure 13.1, KVL gives the following equations for the mesh currents, i_1 and i_2:

   $5i_1 - 2i_2 = 13$
   $-2i_1 + 3i_2 = -8$

   Using determinants, find the value of i_1 and i_2. What current is flowing through the 2 Ω resistor?

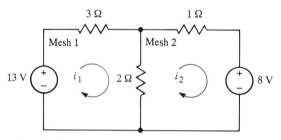

Figure 13.1

Solving for three variables

There are many instances when we have not two but three unknown variables to calculate. For three unknowns, such as x, y and z, we need three equations. Here they are:

$$a_1x + b_1y + c_1z = d_1$$
$$a_2x + b_2y + c_2z = d_2$$
$$a_3x + b_3y + c_3z = d_3$$

If we were working this through step by step, the next stage in the discussion would be to derive a set of equations corresponding to equations (5) and (8) above, only for three variables. Such equations would be even lengthier than (5) and (8) so we will by-pass that stage and go straight to writing out the solutions in determinant form. Let us look at the solution for x:

$$x = \frac{\begin{vmatrix} d_1 & b_1 & c_1 \\ d_2 & b_2 & c_2 \\ d_3 & b_3 & c_3 \end{vmatrix}}{\begin{vmatrix} a_1 & b_1 & c_1 \\ a_2 & b_2 & c_2 \\ a_3 & b_3 & c_3 \end{vmatrix}}$$

We have used the same rules as in the box on page 215, adapting them for a third-order (3 × 3) determinant. The denominator consists of the coefficients of x, y and z written out in rows and columns, just as they are in the set of equations. The numerator is the same *except* that the coefficients for x are replaced by the constants from the right of the equations. The determinants for finding y and z are written out according to the same rules.

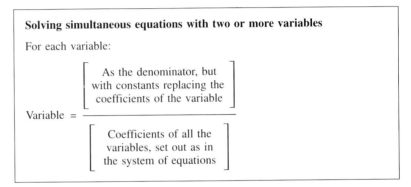

On page 213 we stated that the second-order determinant is a way of writing $a_1b_2 - a_2b_1$. Having omitted the middle stages of the discussion of the third-order determinants, we have not explained what mathematical operations a third-order determinant represents. In other words, we have not said how to evaluate such a determinant.

To understand how to evaluate a determinant of third (or higher) order we need to define two terms, **minor** and **co-factor**. Every element in a determinant has a *minor*, obtained as follows. Given a determinant:

$$\begin{vmatrix} 3 & 2 & 4 \\ 1 & 2 & 5 \\ 7 & 1 & 4 \end{vmatrix}$$

The minor of the 3 at the top left is obtained by crossing out the row and the column which hold the 3, the top row and the first column:

$$\begin{vmatrix} \cancel{3} & \cancel{2} & \cancel{4} \\ \cancel{1} & 2 & 5 \\ \cancel{7} & 1 & 4 \end{vmatrix}$$

This leaves a second-order determinant, the minor of the 3:

$$\begin{vmatrix} 2 & 5 \\ 1 & 4 \end{vmatrix}$$

Similarly, the minor of the 5 is obtained by crossing out the middle row and right column:

$$\begin{vmatrix} 3 & 2 & \cancel{4} \\ \cancel{1} & \cancel{2} & \cancel{5} \\ 7 & 1 & \cancel{4} \end{vmatrix}$$

This gives the minor:

$$\begin{vmatrix} 3 & 2 \\ 7 & 1 \end{vmatrix}$$

Of course, it is not necessary to physically cross out the rows and columns. We *imagine* the crossing out and simply write out the elements which remain.

The co-factor of an element is obtained by multiplying the minor by +1 or −1, according to the position of the element in the determinant. The element at top left is always multiplied by +1 and the pattern of + and − signs is that of a checker-board:

$$\begin{vmatrix} + & - & + & - & + & \ldots \\ - & + & - & + & - & \ldots \\ + & - & + & - & + & \ldots \\ - & + & - & + & - & \ldots \\ + & & & & & \ldots \end{vmatrix}$$

In a third-order determinant, this pattern is:

$$\begin{vmatrix} + & - & + \\ - & + & - \\ + & - & + \end{vmatrix}$$

For the first example above, the 3 at top right, the sign is positive, so the co-factor is:

$$(+1) \begin{vmatrix} 2 & 5 \\ 1 & 4 \end{vmatrix}$$

For the second example, the 5 on the right, the sign is negative, so the co-factor is:

$$(-1) \begin{vmatrix} 3 & 2 \\ 7 & 1 \end{vmatrix}$$

Evaluating a determinant of third (or higher) order

To evaluate a determinant, take the elements of any *one* row or column, multiply each by its co-factor, and then sum these products. Taking the example above and working on the elements in the top row, we obtain:

$$3 \begin{vmatrix} 2 & 5 \\ 1 & 4 \end{vmatrix} \quad -2 \begin{vmatrix} 1 & 5 \\ 7 & 4 \end{vmatrix} \quad +4 \begin{vmatrix} 1 & 2 \\ 7 & 1 \end{vmatrix}$$

> **An alternative technique**
>
> If you find the co-factor routine difficult to remember, here is another way of evaluating a 3×3 (or larger) determinant.
>
> Write out the determinant and repeat the first two or more columns to the right:
>
> $$\begin{vmatrix} 3 & 2 & 4 \\ 1 & 2 & 5 \\ 7 & 1 & 4 \end{vmatrix} \begin{matrix} 3 & 2 \\ 1 & 2 \\ 7 & 1 \end{matrix}$$
>
> Multiply along the diagonals shown, and sum the products:
>
> $$\begin{vmatrix} 3 & 2 & 4 \\ 1 & 2 & 5 \\ 7 & 1 & 4 \end{vmatrix} \begin{matrix} 3 & 2 \\ 1 & 2 \\ 7 & 1 \end{matrix} \quad \begin{matrix} (3 \times 2 \times 4) + (2 \times 5 \times 7) + (4 \times 1 \times 1) \\ = 24 + 70 + 4 = 98 \end{matrix}$$
>
> Then multiply along the opposite diagonal, and sum these products:
>
> $$\begin{vmatrix} 3 & 2 & 4 \\ 1 & 2 & 5 \\ 7 & 1 & 4 \end{vmatrix} \begin{matrix} 3 & 2 \\ 1 & 2 \\ 7 & 1 \end{matrix} \quad \begin{matrix} (7 \times 2 \times 4) + (1 \times 5 \times 3) + (4 \times 1 \times 2) \\ = 56 + 15 + 8 = 79 \end{matrix}$$
>
> Subtract the second sum from the first sum:
>
> $98 - 79 = 19$
>
> The value of the determinant is 19.
>
> Comparing this technique with the co-factor technique, the same numbers are multiplied together and the same products summed (or subtracted) in both. These are just two different ways of setting out the same calculation.

Note the negative sign before the second term, because of the position of the 2 in the determinant. Evaluating the determinants:

$$= 3(8 - 5) - 2(4 - 35) + 4(1 - 14)$$
$$= (3 \times 3) - (2 \times -31) + (4 \times -13)$$
$$= 9 + 62 - 52 = 19$$

If we have a determinant of the fourth order to evaluate, we follow the same routine. In the first stage the four co-factors are third-order determinants, each of which has to be broken down into three second-order determinants. Evaluating determinants of the fourth and higher orders rapidly becomes confusing and we avoid it whenever possible. As we shall see later, there are some ways of simplifying the routines.

Incidentally, to go back to second-order determinants, the way we evaluate these is really the same as for third-order determinants. Take this example:

$$\begin{vmatrix} 4 & 3 \\ 5 & 7 \end{vmatrix}$$

Evaluating along the top row, the co-factor of the 4 is 7. The 3 has a negative position-sign so its co-factor is −5. Summing the element–cofactor products:

$(4 \times 7) + (3 \times -5) = 28 - 15 = 13$

The procedure is the same as that defined on page 213.

Test yourself 13.3

1. Evaluate the determinant on page 217, using the elements of (a) the right column, and (b) the middle row, to confirm that the same result is obtained whatever row or column is used.

2. Evaluate, using (a) any row, and (b) any column, and confirm that both calculations give the same result.

$$\begin{vmatrix} -2 & 1 & 4 \\ 3 & 2 & 5 \\ -1 & 3 & 9 \end{vmatrix}$$

3. Evaluate, using any row or column. Which row or column leads to the easiest calculation?

$$\begin{vmatrix} 3 & 2 & -5 \\ 1 & 4 & 3 \\ 4 & 0 & -1 \end{vmatrix}$$

4. Using the rules in the box, solve these sets of simultaneous equations:

   a $2x + 3y + z = 20$
   $x - y + 4z = 1$
   $3x + 2y - 2z = 14$

   b $2x + 4y - z = -3$
   $3x - 2y - 5z = 1$
   $3y + 4z = 6$

Explore this

Evaluating determinants is such a regular routine that it is easily programmable on a micro. Write a program for evaluating 3×3 determinants and use it to check your working of the exercises.

Making it easier

As questions 3 and 4b of 'Test yourself 13.3' demonstrate, evaluating a third-order determinant is much easier if one of the elements is a zero. It is even easier if two or more elements are zero. When evaluating, always pick a row or column containing one or more zeros, if possible. Unfortunately, practical circuits do not often lead to determinants with zeros in them. We must look for ways of *introducing* zeros into the determinants before we try to evaluate them. To find ways of doing this, we examine some interesting properties of determinants.

Property 1 – common factors

Consider the determinant studied on page 217, which has the value 19. Here is the same determinant, but with the elements in the first column multiplied by 3:

$$\begin{vmatrix} 9 & 2 & 4 \\ 3 & 2 & 5 \\ 21 & 1 & 4 \end{vmatrix}$$

Evaluating this using the middle column:

$$= -2 \begin{vmatrix} 3 & 5 \\ 21 & 4 \end{vmatrix} + 2 \begin{vmatrix} 9 & 4 \\ 21 & 4 \end{vmatrix} - 1 \begin{vmatrix} 9 & 4 \\ 3 & 5 \end{vmatrix}$$

$$= -2(12 - 105) + 2(36 - 84) - 1(45 - 12)$$

$$= (-2 \times -93) + (2 \times -48) - 33$$

$$= 186 - 96 - 33 = 57$$

The value of this determinant is three times the value of the original determinant. By following through the calculation, noting values which are three times what they would otherwise have been, it is easy to see why this result is obtained. The rule is that multiplying any row or column by a constant amount multiplies the value of the determinant by the same constant amount.

The converse also applies, that if the elements of any row or column have a common factor, they can all be divided by that factor, and the factor placed outside the determinant as a multiplier. For example:

$$\begin{vmatrix} 2 & 5 & 4 \\ 6 & 3 & 9 \\ 4 & 5 & 1 \end{vmatrix} = 2 \begin{vmatrix} 1 & 5 & 4 \\ 3 & 3 & 9 \\ 2 & 5 & 1 \end{vmatrix} = 2 \times 3 \begin{vmatrix} 1 & 5 & 4 \\ 1 & 1 & 3 \\ 2 & 5 & 1 \end{vmatrix}$$

The first column has the common factor 2, the middle row has the common factor 3. As a result of these substitutions we have reduced the size of the elements in the first column and middle row, so making the subsequent calculations easier.

Property 2 – combining rows or columns

Taking the same original determinant as an example, we add to each element in the third column, five times the corresponding element of the second column:

$$\begin{vmatrix} 3 & 2 & 4+10 \\ 1 & 2 & 5+10 \\ 7 & 1 & 4+5 \end{vmatrix} = \begin{vmatrix} 3 & 2 & 14 \\ 1 & 2 & 15 \\ 7 & 1 & 9 \end{vmatrix}$$

Evaluating the new determinant by the bottom row:

$$= 7\begin{vmatrix} 2 & 14 \\ 2 & 15 \end{vmatrix} - 1\begin{vmatrix} 3 & 14 \\ 1 & 15 \end{vmatrix} + 9\begin{vmatrix} 3 & 2 \\ 1 & 2 \end{vmatrix}$$

$$= 7(30 - 28) - 1(45 - 14) + 9(6 - 2)$$

$$= 7 \times 2 - 31 + 9 \times 4 = 14 - 31 + 36 = 19$$

The value of the determinant is unaltered. The operation just performed on this determinant seems merely to have made its numbers larger and more difficult to work with. However, if we remember that multiplying each element by a constant also includes multiplying by a *negative* constant there is the possibility of reducing some of the numbers to zero. Suppose that we take the same determinant but this time add to each element in the top row -1 times the corresponding element in the bottom row. In other words, we subtract from the top row the corresponding element in the bottom row, to obtain:

$$\begin{vmatrix} 3-7 & 2-1 & 4-4 \\ 1 & 2 & 5 \\ 7 & 1 & 4 \end{vmatrix} = \begin{vmatrix} -4 & 1 & 0 \\ 1 & 2 & 5 \\ 7 & 1 & 4 \end{vmatrix}$$

With one zero, evaluation involves only two co-factors, but we can do better than this. Add to the first column four times the second column:

$$= \begin{vmatrix} -4+4 & 1 & 0 \\ 1+8 & 2 & 5 \\ 7+4 & 1 & 4 \end{vmatrix} = \begin{vmatrix} 0 & 1 & 0 \\ 9 & 2 & 5 \\ 11 & 1 & 4 \end{vmatrix}$$

Now, operating on the top row, there is only one co-factor to evaluate:

$$= -1\begin{vmatrix} 9 & 5 \\ 11 & 4 \end{vmatrix} = -1(36 - 55) = 19$$

This property is one that is frequently used for simplifying determinants prior to evaluating them. The work saved by reducing elements to zero has to be balanced against the extra work required to obtain the zeros. Being over-ingenious at introducing zeros can be counter-productive.

Property 3 – identical rows or columns

This determinant has two columns that are identical:

$$\begin{vmatrix} 2 & 2 & 5 \\ 3 & 3 & 4 \\ 5 & 5 & 7 \end{vmatrix}$$

Using Property 2, we subtract the second from the first column:

$$\begin{vmatrix} 0 & 2 & 5 \\ 0 & 3 & 4 \\ 0 & 5 & 7 \end{vmatrix}$$

Evaluating by using the first column makes all three co-factors equal to zero, so the determinant as a whole has the value zero. The value of a determinant is unaffected by which row or column is chosen, so whatever way we try, the value remains zero.

The same applies to any determinant for which one row or column is a multiple of another, for example:

$$\begin{vmatrix} 6 & 2 & 5 \\ 9 & 3 & 4 \\ 15 & 5 & 7 \end{vmatrix} = 3 \begin{vmatrix} 2 & 2 & 5 \\ 3 & 3 & 4 \\ 5 & 5 & 7 \end{vmatrix} = 3 \times 0 = 0$$

Using Property 1, we have brought the common factor of the first column out as a multiplier. This leaves the first column identical with the second, as above. The value of the determinant is zero.

It is easy to spot if a determinant has identical rows or columns, or has one row or column a multiple of another. Its value can be taken as zero without any further calculation. This is very helpful when solving differential equations. If a numerator is a zero-valued determinant, the corresponding variable equals zero. If the denominators are zero-valued the solutions to the equation are all indeterminate (page 212).

Property 4 – interchange of rows or columns

Take as an example the determinant of page 217, value 19. Write it out with the second and third columns interchanged and evaluate it (by the third column)

$$\begin{vmatrix} 3 & 4 & 2 \\ 1 & 5 & 2 \\ 7 & 4 & 1 \end{vmatrix} = 2 \begin{vmatrix} 1 & 5 \\ 7 & 4 \end{vmatrix} - 2 \begin{vmatrix} 3 & 4 \\ 7 & 4 \end{vmatrix} + 1 \begin{vmatrix} 3 & 4 \\ 1 & 5 \end{vmatrix}$$

$$= 2(4 - 35) - 2(12 - 28) + 1(15 - 4)$$

$$= 2 \times -31 - 2 \times -16 + 11$$

$$= -62 + 32 + 10 = -19$$

The value is that of the original determinant multiplied by −1. This property applies when any two rows or columns are interchanged. These four properties have been demonstrated with third-order determinants but it can be shown that they apply to determinants of any order.

Test yourself 13.4

Evaluate these determinants.

1. $\begin{vmatrix} 3 & 2 & 3 \\ 1 & -1 & 4 \\ 2 & 3 & 1 \end{vmatrix}$

2. $\begin{vmatrix} 10 & 6 & 1 \\ 2 & 7 & 1 \\ 5 & 3 & 3 \end{vmatrix}$

3. $\begin{vmatrix} 1 & 1 & 1 & -3 \\ 2 & -1 & 3 & 2 \\ 4 & -2 & 3 & 7 \\ 5 & 1 & 2 & 4 \end{vmatrix}$

[Hint: reduce to a third-order matrix by introducing three zeros into a row or column; then proceed to evaluate this third-order matrix.]

4. $\begin{vmatrix} 2 & 7 & 4 \\ 5 & 3 & 10 \\ 3 & -1 & 6 \end{vmatrix}$

5. $\begin{vmatrix} -1 & 3 & -1 \\ 2 & 4 & 0 \\ -3 & 0 & 5 \end{vmatrix}$

Use the method of determinants to solve these sets of simultaneous equations.

6. $3x - 2y + 4z = 2$
   $-3x + y - 2z = 5$
   $3x + 3y + 2z = 7$

7. $4x + 2y - z = -1$
   $3x + 3y + 2z = -11$
   $7x - y - 3z = 22$

8. $2x - 2y + 3z = -9$
   $5x + 3y + z = 10$
   $4x + 5y + 6z = -18$

9. $3w + x + 2y - z = -1$
   $3x + y + 4z = 5$
   $w + 7x - 4y + 3z = 1$
   $-3w - 2x + y + 5z = 12$

10. Applying KVL to the circuit of Figure 13.2a, we obtain the following equations:

    For loop ABCA: $-i_1 + 5i_2 - 3i_3 = 0$
    For loop BCDB: $5i_1 - i_2 - 9i_3 = 0$
    For loop ABDA: $-7i_1 - i_2 + 5i_3 = -1$

    Use the method of determinants to find i_1, i_2 and i_3.

11. A mesh current analysis of the circuit of Figure 13.2b gives the following equations:

    Mesh 1: $8i_1 - 3i_2 - 3i_3 = 4$
    Mesh 2: $-3i_1 + 6i_2 - 3i_3 = 6$
    Mesh 3: $-3i_1 - 3i_2 + 6i_3 = 0$

    Use the method of determinants to find i_1, i_2 and i_3.

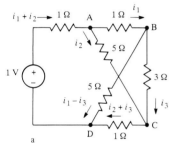

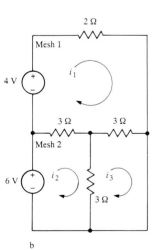

Figure 13.2

Data tables

There are many kinds of data table: results of scientific investigations, a railway timetable, a list of opening hours of a store, or a table of the maximum and minimum daily temperatures and hours of sunshine at a number of holiday resorts. Most of them consist of numbers arranged in rows and columns. Here is an example of a data table, without the headings for the rows and columns:

$$\begin{array}{ccccc} 45 & 50 & 6 & 100 & 22 \\ 20 & 30 & 5 & 100 & 22 \\ 25 & 25 & 5 & 500 & 20 \\ 20 & 40 & 5 & 800 & 14 \end{array}$$

The data in this table refer to various types of transistor. Reading from top to bottom, the rows refer to the BC107, BC108, ZTX300 and 2N3760. Reading from left to right, the columns refer to maximum values of V_{CEO}, V_{CBO} and V_{EBO} in volts, maximum I_C in milliamperes, and the price per transistor, in pence. An array of data arranged like this, in rows and columns, is referred to as a **matrix**. In mathematics we enclose a matrix in brackets, either square or round:

$$\begin{bmatrix} 45 & 50 & 6 & 100 & 22 \\ 20 & 30 & 5 & 100 & 22 \\ 25 & 25 & 5 & 500 & 20 \\ 20 & 40 & 5 & 800 & 14 \end{bmatrix}$$

At first glance a matrix might be mistaken for a determinant, but it is quite different. A determinant is necessarily **square**, having an equal number of rows and columns, whereas a matrix need not be square, as in the example above with four rows and five columns. More significant is the fact that a determinant has a numerical *value*, obtainable by the routines described on pages 213–223. A matrix does not have a value. It would not make sense to think of the timetable of the trains from London to Edinburgh as being replaceable by a single number.

Describing a matrix

The **order** of a matrix is given by the number of rows followed by the number of columns. The example above is of order 4×5. Remember the RC rule: the number of **R**ows comes fi**R**st, the number of **C**olumns comes se**C**ond. In general terms we denote the number of rows by m and the number of columns by n, so we speak of an $m \times n$ matrix. For a **square matrix** we have $m = n$. A matrix may have only one row. For example:

$$\begin{bmatrix} 2 & 5 & 53 & -4 & 7 \end{bmatrix}$$

is a 1×5 matrix. This type of matrix is called a **row matrix** or **row vector**. (The use of matrices to describe vectors is demonstrated on page 28). Similarly a matrix may have only 1 column. For example:

$$\begin{bmatrix} 4 \\ 1 \\ 0 \\ -2 \end{bmatrix}$$

is a 4 × 1 matrix. This is called a **column matrix** or **column vector**. To save space on the page, a column matrix may be printed horizontally in curly brackets:

$$\{4 \quad 1 \quad 0 \quad -2\}$$

It is conventional to represent the elements of a matrix by the letter a followed by a double subscript to indicate the row and column. Thus an $m \times n$ matrix is made up of elements according to this pattern:

$$\begin{bmatrix} a_{11} & a_{12} & a_{13} & \cdots & a_{1n} \\ a_{21} & a_{22} & a_{23} & \cdots & a_{2n} \\ a_{31} & a_{32} & a_{33} & \cdots & a_{3n} \\ \cdots & \cdots & \cdots & \cdots & \cdots \\ a_{m1} & a_{m2} & a_{m3} & \cdots & a_{mn} \end{bmatrix}$$

It is often inconvenient to write out all the terms of a matrix, especially when the same matrix has to be represented several times at different stages of a calculation. One shorthand way of representing a matrix is by enclosing a typical term in brackets. The matrix above is more briefly represented by:

$$[a_{ij}]$$

This indicates that it consists of an array of elements and that we specify any particular element by its row i and its column j. Another even shorter notation is to represent the matrix by a capital letter printed in bold type. In this notation, the matrix above might be represented by:

A

In handwritten text we indicate bold type by underlining with a wavy line:

A̰

In this system a column matrix is represented by a lower-case (small) letter in bold type. For example:

$$\begin{bmatrix} b_{11} \\ b_{21} \\ b_{31} \\ b_{41} \end{bmatrix} \quad \text{is represented by} \quad \mathbf{b}$$

Working with matrices

Although a matrix cannot be evaluated as can a determinant, there are several arithmetical operations that can be performed on it. In these operations the matrix is treated as if it were an ordinary variable. The operations of **matrix arithmetic** include:

Addition: each element of one matrix is added to the corresponding element of another matrix. For example:

$$\begin{bmatrix} 3 & 4 \\ 5 & -1 \end{bmatrix} + \begin{bmatrix} 2 & 3 \\ 4 & 7 \end{bmatrix} = \begin{bmatrix} 5 & 7 \\ 9 & 6 \end{bmatrix}$$

For matrix addition to be possible, both matrices must be of the same order. Note that the result of adding one matrix to another is *another matrix*, of the same order.

Scalar multiplication: each element of a matrix is multiplied by the same factor. For example:

$$3 \times \begin{bmatrix} 3 & 2 \\ -1 & 4 \end{bmatrix} = \begin{bmatrix} 9 & 6 \\ -3 & 12 \end{bmatrix}$$

The product is *another matrix*, of the same order.

Matrix multiplication: one matrix is multiplied by another matrix to produce a third matrix. As always, the result of the calculation is not a value but *another matrix*.

We will not discuss matrix addition or scalar multiplication further, since these operations are not required for solving simultaneous equations, the subject of this chapter. But matrix multiplication is important in this context, so we examine it in more detail below.

Matrix multiplication

This practical example shows how matrix multiplication works and what kinds of problem it can help to solve. There are three different display panels, consisting of light emitting diodes (LEDs). The diodes are red or yellow and each panel has different numbers of each colour. This matrix shows how many LEDs of each colour are mounted on each panel:

	Red	Yellow
Panel A	6	2
Panel B	4	1
Panel C	3	4

We refer to this matrix as **N**, as it lists numbers of LEDs.

To ensure equal brightness, the LEDs of the two colours are made so that they pass different amounts of current:

	Current (mA)
Red	10
Yellow	15

Call this matrix **c**, for current. The small letter indicates that it is a column matrix.

We are to calculate how much current is required for each display when all the LEDs are illuminated. The calculation is easy, using ordinary arithmetic:

Display A: $(6 \times 10) + (2 \times 15) = 60 + 30 = 90$
Display B: $(4 \times 10) + (1 \times 15) = 40 + 15 = 55$
Display C: $(3 \times 10) + (4 \times 15) = 30 + 45 = 75$

We set out the results in a third matrix:

$$\begin{array}{c} \text{Total current} \\ \text{(mA)} \end{array}$$

$$\begin{array}{c} \text{Display A} \\ \text{Display B} \\ \text{Display C} \end{array} \begin{bmatrix} 90 \\ 55 \\ 75 \end{bmatrix}$$

Call this matrix **t**, for total current. This, too, is a column matrix.

Look again at the calculations of the total current and see how they relate to the contents of the matrices. To calculate the total for each display we take the elements from the first *row* of matrix **N**, in order from left to right, and multiply each by the elements from the one (and only) *column* of matrix **c**, in order from top to bottom. Multiplying the elements of the *rows* of one matrix by the elements of the *columns* of a second matrix is called the **matrix multiplication**.

We symbolize the operation by an equation:

$$\begin{bmatrix} 6 & 2 \\ 4 & 1 \\ 3 & 4 \end{bmatrix} \times \begin{bmatrix} 10 \\ 15 \end{bmatrix} = \begin{bmatrix} 90 \\ 55 \\ 75 \end{bmatrix}$$

or even more briefly by:

N × c = t

See how the short form of the equation states the basic multiplication carried out between pairs of elements:

the numbers of LEDs × current per LED = total current

An important point about the calculation above is the number of rows and columns in the various matrices:

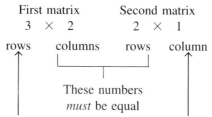

The numbers of rows and columns in
the result are the same as these

These rules apply to all matrix multiplication.

Matrix multiplication, 2 × 2 matrices

$$\begin{bmatrix} a & b \\ c & d \end{bmatrix} \times \begin{bmatrix} e & f \\ g & h \end{bmatrix} = \begin{bmatrix} ae + bg & af + bh \\ ce + dg & cf + dh \end{bmatrix}$$

> **Matrix multiplication, any order**
>
> 1 The number of columns in the first matrix must equal the number of rows in the second matrix. If not, multiplication is impossible.
> 2 Set out a blank product matrix with the same number of rows as the first matrix and the same number of columns as the second matrix.
> 3 Take each row of the first matrix, with each column of the second matrix:
>
> $$\begin{vmatrix} r_1 & r_2 & r_3 & \ldots & r_n \end{vmatrix} \times \begin{vmatrix} c_1 \\ c_2 \\ c_3 \\ \ldots \\ c_n \end{vmatrix}$$
>
> 4 Sum the products:
>
> $$r_1 c_1 + r_2 c_2 + r_3 c_3 + \ldots r_n c_n = S$$
>
> 5 Write the result S in the product matrix at the location where the row and column cross:
>
> $$\begin{bmatrix} - & - & - & - \\ - & S & - & - \\ - & - & & \\ - & & & \end{bmatrix}$$
>
> 6 Do this for every row by every column.

Test yourself 13.5

If

$$A = \begin{bmatrix} 2 & 1 \\ 3 & -1 \end{bmatrix} \qquad b = \begin{bmatrix} 3 \\ 4 \end{bmatrix} \qquad c = \begin{bmatrix} -2 & 1 \end{bmatrix}$$

$$D = \begin{bmatrix} 2 & 0 & 1 \\ 3 & 1 & -1 \\ 2 & 2 & 3 \end{bmatrix} \qquad E = \begin{bmatrix} 1 & 2 \\ 0 & -2 \\ 3 & 1 \end{bmatrix}$$

calculate the following matrix products, if possible:

Ab, Ac, bc, cb, A^2, bD, cA, DE, ED, EA, AE.

Order of multiplication   Your answers to the questions above will have shown that the order in which two matrices are to be multiplied is important. When we multiply two numbers or variables together, the order in which they are written makes no difference to the result. For example:

$3 \times 7 = 21$
$7 \times 3 = 21$

Similarly:

$$a \times b = ab$$
$$b \times a = ab$$

We say that multiplication of real numbers is **commutative**.

In the example of the LEDs we multiplied **N** by **c**. It is clear that it is impossible to multiply **c** by **N** because **c** has only one column but **N** has three rows. It is not possible to find pairs of elements to multiply together as the rules require. As we have already shown, the number of columns in the first matrix *must* equal the number of rows in the second matrix.

With two *square* matrices, it is always possible to find pairs of elements to multiply, in whichever order we multiply them. See what happens in this example with two 2 × 2 matrices:

$$\begin{bmatrix} 2 & 5 \\ 3 & 1 \end{bmatrix} \times \begin{bmatrix} 4 & 1 \\ 2 & 6 \end{bmatrix} = \begin{bmatrix} 2\times4+5\times2 & 2\times1+5\times6 \\ 3\times4+1\times2 & 3\times1+1\times6 \end{bmatrix} = \begin{bmatrix} 8 & 32 \\ 14 & 9 \end{bmatrix}$$

Now multiplying them in the other order:

$$\begin{bmatrix} 4 & 1 \\ 2 & 6 \end{bmatrix} \times \begin{bmatrix} 2 & 5 \\ 3 & 1 \end{bmatrix} = \begin{bmatrix} 4\times2+1\times3 & 4\times5+1\times1 \\ 2\times2+3\times6 & 2\times5+6\times1 \end{bmatrix} = \begin{bmatrix} 11 & 21 \\ 22 & 16 \end{bmatrix}$$

The elements pair differently in the second case, giving a different result. In almost all cases, multiplying **A** by **B** and multiplying **B** by **A** gives different results. Matrix multiplication is not commutative.

Identity matrix

In ordinary (scalar) multiplication we know that:

$$4 \times 1 = 4$$

Multiplying anything by 1 has no effect on its value. We say that 1 is the **identity operator** for scalar multiplication. Here is an example of the same thing, with matrices:

$$\begin{bmatrix} 2 & 5 \\ 3 & 1 \end{bmatrix} \times \begin{bmatrix} 1 & 0 \\ 0 & 1 \end{bmatrix} = \begin{bmatrix} 2\times1+5\times0 & 2\times0+5\times1 \\ 3\times1+1\times0 & 3\times0+1\times1 \end{bmatrix} = \begin{bmatrix} 2 & 5 \\ 3 & 1 \end{bmatrix}$$

Multiplication by $\begin{bmatrix} 1 & 0 \\ 0 & 1 \end{bmatrix}$ has left the matrix unchanged.

We call $\begin{bmatrix} 1 & 0 \\ 0 & 1 \end{bmatrix}$ the **identity matrix** for 2 × 2 matrix multiplication.

It is given the special symbol **I**. In symbols, for any matrix:

A . I = A

If we multiply in the reverse order:

$$\begin{bmatrix} 1 & 0 \\ 0 & 1 \end{bmatrix} \times \begin{bmatrix} 2 & 5 \\ 3 & 1 \end{bmatrix} = \begin{bmatrix} 1\times2+0\times3 & 1\times5+0\times1 \\ 0\times2+1\times3 & 0\times5+1\times1 \end{bmatrix} = \begin{bmatrix} 2 & 5 \\ 3 & 1 \end{bmatrix}$$

In symbols:

I . A = A

As an exception to the general rules, multiplication by the identity matrix is commutative. Identity matrices for 3 × 3 and square matrices of higher order are constructed on the same principle. The elements of the leading diagonal are 1, and the remainder are 0.

The LED display problem again

In the example on page 227 we are told the numbers of LEDs in each display and the amount of current each LED requires. We are asked to calculate the total current for each type of LED. The matrix equation for solving this problem is:

$$\mathbf{t} = \mathbf{Nc} \qquad (9)$$

where the matrices **t**, **N** and **c** are as defined on page 227.

Now to look at a different problem. Suppose that we are told how many LEDs of each colour are mounted in each kind of display, and the total current for each display. We are asked how much current the LEDs of each colour require. In other words, we are given **N** and **t** and have to calculate **c**. This problem can be solved by setting out a system of simultaneous equations:

$$6c_1 + 2c_2 = 90 \qquad (10)$$
$$4c_1 + c_2 = 55 \qquad (11)$$
$$3c_1 + 4c_2 = 75 \qquad (12)$$

in which c_1 and c_2 are the currents taken by red LEDs and yellow LEDs respectively. Since there are two unknowns, c_1 and c_2, we actually only need two equations. We will use equations (10) and (11). However, instead of solving these by elimination or by using determinants, we will solve them by using matrices. Since we are ignoring equation (12), we re-define **N** and **t** with only two rows each:

$$\mathbf{N} = \begin{bmatrix} 6 & 2 \\ 4 & 1 \end{bmatrix} \qquad \mathbf{t} = \begin{bmatrix} 90 \\ 55 \end{bmatrix}$$

Now that we have the appropriate matrices, the problem can be solved by changing the subject of equation (10):

$$\mathbf{c} = \frac{\mathbf{t}}{\mathbf{N}} \qquad (13)$$

The difficulty with this is that there is no operation for division of matrices. Division of *determinants* is possible because we simply evaluate the determinants and divide one number by the other number. The result is a *number*. But, with matrices, the result must be a *matrix*. In this example we have to find the matrix **c**.

The difficulty is resolved by converting equation (13) into a matrix *multiplication*, for which we already have a set of rules. With ordinary numbers, the operation of dividing by n is identical to the operation of multiplying by $1/n$. Instead of dividing by a number, we multiply by its *reciprocal* (page 23) or *inverse*.

The matrix equation (13) is rewritten:

$$c = N^{-1}t$$

where N^{-1} is the inverse of N. Now we must find out what we mean by the inverse of a matrix.

The inverse of a 2 × 2 matrix

The routine is:

(1) Exchange the elements on the leading diagonal.
(2) Change the signs of the elements on the other diagonal.
(3) Divide by the determinant.

Trying this on N we obtain:

$$N^{-1} = \frac{\begin{bmatrix} 1 & -2 \\ -4 & 6 \end{bmatrix}}{-2}$$

Check this to see if it really is the inverse, if N^{-1} really does what an inverse is expected to do. When we multiply a number n by its inverse n^{-1} the rule of indices (page 23) tells us that the result is n^0, which is equal to 1. If we multiply the matrix A by its inverse A^{-1} we should therefore obtain I, the identity matrix. Let us confirm this with the example above:

$$A \cdot A^{-1} = \begin{bmatrix} 6 & 2 \\ 4 & 1 \end{bmatrix} \times \frac{\begin{bmatrix} 1 & -2 \\ -4 & 6 \end{bmatrix}}{-2} = \frac{\begin{bmatrix} -2 & 0 \\ 0 & -2 \end{bmatrix}}{-2} = \begin{bmatrix} 1 & 0 \\ 0 & 1 \end{bmatrix} = I$$

Having found the inverse matrix, we use it to solve the problem:

$$c = N^{-1} \cdot t = \frac{\begin{bmatrix} 1 & -2 \\ -4 & 6 \end{bmatrix}}{-2} \times \begin{bmatrix} 90 \\ 55 \end{bmatrix} = \frac{\begin{bmatrix} 1 \times 90 + -2 \times 55 \\ -4 \times 90 + 6 \times 55 \end{bmatrix}}{-2}$$

$$= \frac{\begin{bmatrix} -20 \\ -30 \end{bmatrix}}{-2} = \begin{bmatrix} 10 \\ 15 \end{bmatrix}$$

$c_1 = 10$ and $c_2 = 15$.

These, of course, are the same values as we used in the original problem. Note how the matrix method produces a matrix which holds *all* the unknowns. Compare this with the determinant method in which we have to perform a separate calculation for each unknown.

Finding the inverse of a 3×3 matrix

It is not often that we would use the matrix method for solving such a simple problem as the LED problem, which has only two unknowns. More often we shall have at least three unknowns with an equal number of equations. We need a method for finding the inverse of a matrix of 3 × 3 or greater order. The method described here applies to any order. When you have seen how it works, it is clear that the method for 2 × 2 matrices is really the same method simplified, because there are only two rows and two columns.

Remember that an inverse can be found only for a *square* matrix (which was why we dropped the redundant equation (12) in the problem and redefined the matrices). Below we describe *how* to find the inverse of a square matrix of any size, and then demonstrate that the result we obtain has the property required of an inverse. As before, we refer to the matrix as **A** and its inverse as $\mathbf{A}^{-1}$.

Step 1: Calculate $|\mathbf{A}|$, the determinant of **A**. If $|\mathbf{A}| = 0$, there is no inverse and we abandon the calculation.

Step 2: Calculate the **co-factor matrix**. This is a matrix of the same order as **A** in which every element of **A** is replaced by its co-factor (page 217). For example, if:

$$\mathbf{A} = \begin{bmatrix} 4 & 1 & 6 \\ 2 & 3 & -1 \\ 3 & 2 & 5 \end{bmatrix}$$

the co-factor of the 4 is $(3 \times 5) - (-1 \times 2) = 17$; and the co-factor of the 1 is $-[(2 \times 5) - (-1 \times 3)] = -13$ (the negative sign in front is the position sign, page 218).

Other co-factors are calculated in the same way and the co-factor matrix is:

$$\begin{bmatrix} 17 & -13 & -5 \\ 7 & 2 & -5 \\ -19 & 16 & 10 \end{bmatrix}$$

Step 3: Form the **transpose** of the co-factor matrix, by interchanging its rows and columns. The result is known as the **adjoint matrix**:

$$\text{Adj } \mathbf{A} = \begin{bmatrix} 17 & 7 & -19 \\ -13 & 2 & 16 \\ -5 & -5 & 10 \end{bmatrix}$$

Step 4: The inverse is the adjoint matrix divided by the determinant. The determinant, $|\mathbf{A}| = 25$, so:

$$\mathbf{A}^{-1} = \frac{\text{Adj } \mathbf{A}}{|\mathbf{A}|} = \frac{\begin{bmatrix} 17 & 7 & -19 \\ -13 & 2 & 16 \\ -5 & -5 & 10 \end{bmatrix}}{25}$$

Testing the inverse

Check the inverse matrix as before:

$$\mathbf{A} \cdot \mathbf{A}^{-1} = \begin{bmatrix} 4 & 1 & 6 \\ 2 & 3 & -1 \\ 3 & 2 & 5 \end{bmatrix} \times \frac{\begin{bmatrix} 17 & 7 & -19 \\ -13 & 2 & 16 \\ -5 & -5 & 10 \end{bmatrix}}{25}$$

$$= \frac{\begin{bmatrix} 25 & 0 & 0 \\ 0 & 25 & 0 \\ 0 & 0 & 25 \end{bmatrix}}{25} = \begin{bmatrix} 1 & 0 & 0 \\ 0 & 1 & 0 \\ 0 & 0 & 1 \end{bmatrix} = \mathbf{I}$$

The result is the identity matrix, confirming that the method does give $\mathbf{A}^{-1}$.

Another inverse

As a quick summary of the method, here is another example:

$$\mathbf{A} = \begin{bmatrix} 2 & 2 & 1 \\ -1 & 3 & 0 \\ 4 & -2 & 3 \end{bmatrix}$$

$$|\mathbf{A}| = 14$$

$$\text{co-factor matrix} = \begin{bmatrix} 9 & 3 & -10 \\ -8 & 2 & 12 \\ -3 & -1 & 8 \end{bmatrix}$$

$$\text{Adj } \mathbf{A} = \begin{bmatrix} 9 & -8 & -3 \\ 3 & 2 & -1 \\ -10 & 12 & 8 \end{bmatrix}$$

$$\mathbf{A}^{-1} = \frac{\begin{bmatrix} 9 & -8 & -3 \\ 3 & 2 & -1 \\ -10 & 12 & 8 \end{bmatrix}}{14}$$

Test yourself 13.6

Find the inverses of the following matrixes, if possible:

1. $\begin{bmatrix} 4 & 5 \\ 3 & 7 \end{bmatrix}$
2. $\begin{bmatrix} 2 & -5 \\ 3 & 1 \end{bmatrix}$
3. $\begin{bmatrix} 6 & 3 \\ 4 & 2 \end{bmatrix}$
4. $\begin{bmatrix} -3 & 5 \\ -2 & 7 \end{bmatrix}$
5. $\begin{bmatrix} 0 & 1 & 1 \\ 1 & 0 & 1 \\ 1 & 1 & 0 \end{bmatrix}$
6. $\begin{bmatrix} 1 & 1 & 2 \\ 2 & 1 & 1 \\ 1 & 2 & 2 \end{bmatrix}$

$$7 \begin{bmatrix} 3 & 2 & 4 \\ -1 & -2 & 0 \\ 1 & -3 & 5 \end{bmatrix} \quad 8 \begin{bmatrix} 3 & 0 & -1 \\ 2 & 2 & -4 \\ -1 & 1 & 3 \end{bmatrix} \quad 9 \begin{bmatrix} 1 & -1 & 0 \\ -2 & 3 & 2 \\ 3 & -2 & 1 \end{bmatrix}$$

Another problem

This example shows how the matrix method is used for solving a system of three simultaneous equations for three unknowns, x_1, x_2 and x_3:

$$3x_1 + 2x_2 - 2x_3 = 4$$
$$-2x_1 + 3x_2 - 5x_3 = 15$$
$$2x_1 - x_2 + 2x_3 = -7$$

Putting this in matrix form, we need matrix **A** for the coefficients, matrix **x** for the unknown variables, and matrix **k** for the constants on the right sides of the equations:

$$\mathbf{A} = \begin{bmatrix} 3 & 2 & -2 \\ -2 & 3 & -5 \\ 2 & -1 & 2 \end{bmatrix} \quad \mathbf{x} = \begin{bmatrix} x_1 \\ x_2 \\ x_3 \end{bmatrix} \quad \mathbf{k} = \begin{bmatrix} 4 \\ 15 \\ -7 \end{bmatrix}$$

The matrix equation is: $\mathbf{A} \cdot \mathbf{x} = \mathbf{k}$

And the solution is: $\mathbf{x} = \mathbf{A}^{-1} \cdot \mathbf{k}$

Using the method just described for finding the inverse gives:

$$\mathbf{A}^{-1} = \frac{\begin{bmatrix} 1 & -2 & -4 \\ -6 & 10 & 19 \\ -4 & 7 & 13 \end{bmatrix}}{-1}$$

The solution to the equations is:

$$\mathbf{x} = \frac{\begin{bmatrix} 1 & -2 & -4 \\ -6 & 10 & 19 \\ -4 & 7 & 13 \end{bmatrix}}{-1} \times \begin{bmatrix} 4 \\ 15 \\ -7 \end{bmatrix} = \frac{\begin{bmatrix} 2 \\ -7 \\ -2 \end{bmatrix}}{-1} = \begin{bmatrix} -2 \\ 7 \\ 2 \end{bmatrix}$$

$x_1 = -2$, $x_2 = 7$, and $x_3 = 2$. As before, a single matrix calculation provides one matrix that contains *all* the unknown variables.

Test yourself 13.7

Solve these systems of equations, using the matrix method:

1 $3x_1 - x_2 - 2x_3 = 4$
   $x_1 + 3x_2 - x_3 = 8$
   $2x_1 + 2x_2 - 3x_3 = 2$

2 $3x_1 + x_2 - 2x_3 = 15$
$x_1 + 3x_2 - x_3 = 14$
$2x_1 - 3x_3 = 13$

3 $x_1 + 2x_2 + 3x_3 = 0$
$2x_1 + x_2 - 4x_3 = 10$
$x_1 - x_2 + 2x_3 = 1$

If you need further practice, try solving some of the equations of earlier exercises, using the matrix method.

Row equivalents

Another technique for solving simultaneous equations involves taking the whole equations into one matrix. Given the equations:

$$3x_1 - x_2 = 9$$
$$4x_1 + 3x_2 = -1$$

the corresponding matrix is:

$$\begin{bmatrix} 3 & -1 & 9 \\ 4 & 3 & -1 \end{bmatrix}$$

Since this consists of the coefficient matrix combined with the constant matrix, it is called an **augmented matrix**. In this method we do not work with matrix equations, but just with this matrix alone. When we eventually find the solutions, these consist of two equations:

$$x_1 = a$$
$$x_2 = b$$

where a and b are the as yet undiscovered values of x_1 and x_2. Given that the coefficients of x_1 and x_2 are both 1, these solution equations can also be written as an augmented matrix:

$$\begin{bmatrix} 1 & 0 & a \\ 0 & 1 & b \end{bmatrix}$$

The aim of this method is to start with the augmented matrix of the original equation and to convert this into the augmented matrix of the solution, with 1's and 0's where shown. Then the values of x_1 and x_2 will be at a and b.

Note that in this method the matrix is a *list* of equations. Each row corresponds to one equation. This being so, there are various operations that can be carried out on the *rows* (not the columns) without affecting the solution.

- Rows can be interchanged; listing the equations in a different order is not going to affect the solution.

- A row can be multiplied throughout by a non-zero constant; multiplying the whole of both sides of an equation has no effect on the solution (page 36).

- A multiple of one row can be added to another row; this is equivalent to adding the same amount to both sides of an equation, and has no effect on the solution (page 35).

Remember that everything that is done has to be done to the *whole* of the row, otherwise the equation it represents will become unbalanced and a false result obtained.

The task is to find a way of using the operations listed above to convert the original matrix into the solution matrix. There is often more than one path to follow. This is one possible way:

Original matrix is
$$\begin{bmatrix} 3 & -1 & 9 \\ 4 & 3 & -1 \end{bmatrix}$$ Row (1) Row (2)

Multiply (1) by 1/3
$$\begin{bmatrix} 1 & -1/3 & 3 \\ 4 & 3 & -1 \end{bmatrix}$$

Subtract 4 times (1) from (2)
$$\begin{bmatrix} 1 & -1/3 & -3 \\ 0 & 13/3 & -13 \end{bmatrix}$$

Multiply (2) by 3/13
$$\begin{bmatrix} 1 & -1/3 & 3 \\ 0 & 1 & -3 \end{bmatrix}$$

Add 1/3 of (2) to (1)
$$\begin{bmatrix} 1 & 0 & 2 \\ 0 & 1 & -3 \end{bmatrix}$$

This is the solution matrix:

$x_1 = 2, \quad x_2 = -3.$

The general strategy is to try to place a 1 at the left of the top row, perhaps exchanging rows to do so. Then try to fill the remainder of the 1st column with zeros. This is done by subtracting multiples of the top row from the other row. The fact that there is a 1 in the first column of the first row makes this easy. This process is repeated, placing a 1 in the second column of the second row, and filling the remainder of that column with zeros. Repeat this until the solution matrix is obtained.

Here is an example with three equations:

$2x_1 + x_2 - 2x_3 = -1$

$x_1 - 3x_2 - 3x_3 = 8$

$3x_1 - x_2 + 2x_3 = 11$

The augmented matrix is:
$$\begin{bmatrix} 2 & 1 & -2 & -1 \\ 1 & -3 & -3 & 8 \\ 3 & -1 & 2 & 11 \end{bmatrix} \begin{matrix} \text{Row (1)} \\ \text{Row (2)} \\ \text{Row (3)} \end{matrix}$$

Exchange (1) and (2)
$$\begin{bmatrix} 1 & -3 & -3 & 8 \\ 2 & 1 & -2 & -1 \\ 3 & -1 & 2 & 11 \end{bmatrix}$$

Subtract two times (1) from (2)
$$\begin{bmatrix} 1 & -3 & -3 & 8 \\ 0 & 7 & 4 & -17 \\ 3 & -1 & 2 & 11 \end{bmatrix}$$

Subtract three times (1) from (3)
$$\begin{bmatrix} 1 & -3 & -3 & 8 \\ 0 & 7 & 4 & -17 \\ 0 & 8 & 11 & -13 \end{bmatrix}$$

Subtract (2) from (3)
$$\begin{bmatrix} 1 & -3 & -3 & 8 \\ 0 & 7 & 4 & -17 \\ 0 & 1 & 7 & 4 \end{bmatrix}$$

Exchange (2) and (3)
$$\begin{bmatrix} 1 & -3 & -3 & 8 \\ 0 & 1 & 7 & 4 \\ 0 & 7 & 4 & -17 \end{bmatrix}$$

Subtract seven times (2) from (3)
$$\begin{bmatrix} 1 & -3 & -3 & 8 \\ 0 & 1 & 7 & 4 \\ 0 & 0 & -45 & -45 \end{bmatrix}$$

Multiply (3) by $-1/45$
$$\begin{bmatrix} 1 & -3 & -3 & 8 \\ 0 & 1 & 7 & 4 \\ 0 & 0 & 1 & 1 \end{bmatrix}$$

Add three times (2) to (1)
$$\begin{bmatrix} 1 & 0 & 18 & 20 \\ 0 & 1 & 7 & 4 \\ 0 & 0 & 1 & 1 \end{bmatrix}$$

Subtract eighteen times (3) from (1)
$$\begin{bmatrix} 1 & 0 & 0 & 2 \\ 0 & 1 & 7 & 4 \\ 0 & 0 & 1 & 1 \end{bmatrix}$$

Subtract seven times (3) from (2)
$$\begin{bmatrix} 1 & 0 & 0 & 2 \\ 0 & 1 & 0 & -3 \\ 0 & 0 & 1 & 1 \end{bmatrix} \begin{matrix} \text{Solution} \\ \text{matrix} \end{matrix}$$

The solution matrix has 0's and 1's in the correct format in the first three columns; the last column holds the solutions:

$x_1 = 2$, $x_2 = -3$ and $x_3 = 1$.

It almost seems that we have come full circle from the elimination method of page 41. The method described above is really nothing more than a way of manipulating equations without having to write them out in full. Like the elimination method, it requires a certain amount of knack to spot the best way of working through to the solution. The main advantage is that, having the figures set out in a neat array, and having a few clear rules to follow, the augmented matrix method makes it easier and quicker to arrive at a solution.

Test yourself 13.8

Use the augmented matrix method to solve some of the systems of equations in previous exercises in this chapter.

14 Beyond the number line

> The maths in this chapter is used for describing alternating currents and voltages, particularly when represented by phasors. It is also important in studying magnetically coupled circuits and resonant circuits and in calculating the frequency response of filters and related circuits.
>
> You need to know about numbers and the number line (page 3), indices (page 23), logarithms (page 25), vectors (page 28), quadratic equations (page 39), rectangular and polar coordinates (page 93), and sines and cosines (page 65).

Every real number has its place on the number line (Figure 14.1). The line is divided into two regions:

- numbers less than zero (negative numbers)
- numbers equal to or greater than zero (positive numbers)

When *any* real number is squared, the square is positive. This is the rule of signs when multiplying:

positive × positive ⇒ positive
negative × negative ⇒ positive

There is no way in which the square of a real number can be negative.

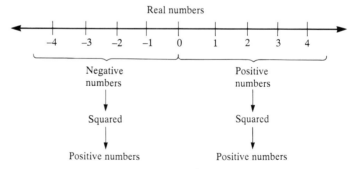

Figure 14.1

Using the imagination

Although all real numbers have positive squares, it is interesting to imagine a number which has a *negative* square, and to deduce what the properties of such a number might be. As the basic **imaginary number** we select the number which, when squared, equals −1. Since it has no value that is expressible in real numbers, we represent it by a symbol. Mathematicians use the symbol i, because i and 1 look rather alike. The difficulty with this is that, in electronics and the theory of electric circuits, we commonly use *i* for representing current. This is why electronic and electrical engineers use the symbol j instead of *i*. We define j by the equation:

$$j = \sqrt{-1} \quad \text{or} \quad j^2 + 1 = 0$$

As we shall see later, j is rather more than just an imaginary number. It is an operator which performs real actions. For the moment, we will treat it as if j were just a number.

Working with j

It is not difficult to decide how j behaves when it is subjected to some of the simpler arithmetical operations.

Addition

It is clear that: $\quad \sqrt{-1} + \sqrt{-1} = 2\sqrt{-1}$

In terms of j: $\quad j + j = j2$

Note that we use the form j2, not 2j. This is because j is not just a quantity. As explained later, it is an *operator* and therefore is written *before* the number on which it operates.

Similarly: $\quad j2 + j = j3$

In general: $\quad ja + jb = j(a + b)$

This leads to the rule that, when two imaginary numbers are added together, the sum is also imaginary.

Multiplication by a real number

Multiplication follows the rules of algebra:

$$j2 \times 3 = j6$$

When an imaginary number is multiplied by a real number, the product is imaginary.

Multiplication by another imaginary number

This gives results that are unlike those obtained with real numbers. By definition:

$$j \times j = \sqrt{-1} \times \sqrt{-1} = -1$$

For larger imaginary numbers, we collect together the imaginary numbers and the real numbers:

$$j2 \times j3 = (j \times j)(2 \times 3) = -1 \times 6 = -6$$

In general: $\quad ja \times jb = -ab$

When two imaginary numbers are multiplied together, the product is real.

Powers of j

It follows from the previous paragraph, as well as from the definition of j, that:

$$j^2 = j \times j = -1$$

By extension of this equation, and by the index rules (page 24):

$$j^3 = j \times j^2 = j \times -1 = -j$$

and
$$j^4 = j \times j^3 = j \times -j = -(-1) = 1$$

After this, with increasing powers of j, the sequence repeats:

$$j^5 = j, \quad j^6 = -1, \quad j^7 = -j, \quad \text{and so on.}$$

The rule for finding the value of a power of j is to divide the power by 4, and note the remainder:

If the remainder is 0, the result is 1
1, j
2, −1
3, −j

It seems that, though it is imaginary, j has properties that are easily understood and easy to work with. Also, though it often follows the rules of ordinary algebra, it also has some distinctive behaviour of its own.

Test yourself 14.1

1 Which numbers in this list are imaginary?

3, j, j2, −5, π, $\sqrt{-7}$, −j6, e, ⅔, −j4, $\sqrt{2}$, jπ

Simplify these expressions.

2 j2 + j5   3 j4 − j3
4 j3 × 2   5 4 × j7
6 j4 × j3   7 j^{11}
8 j2 × −j3   9 $j^3 2 \times j^6 5$
10 j2 − $j^2 3 \times j^3 7$

The imaginary number line

The discussion above shows that imaginary numbers can be positive or negative and have a range of values. They can be represented by an *imaginary number line* (Figure 14.2). For each number on the real number line, there is

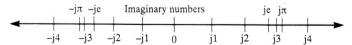

Figure 14.2

a corresponding number on the imaginary number line. For example, the real numbers:

3, −5, 0, 4.27, π, ¾ and $\sqrt{7}$

all have their counterpart on the imaginary number line:

j3, −j5, j0 (= 0), j4.27, jπ, j¾, and j$\sqrt{7}$

The imaginary number line is just like the real number line, except that every number on it is a real number operated on by j. The real number line is continuous from −∞ to +∞; the imaginary number line is continuous from −j∞ to +j∞.

There is one number that is found on *both* lines. That number is zero, for j0 = 0. If we draw both lines on a single diagram, and let them cross at zero, we have a way of representing both real and imaginary numbers on the same diagram (Figure 14.3). The most interesting thing about this diagram is the area around the lines. Any point in this area can be represented by a pair of values that indicate its position. The pair of values is like the rectangular coordinates described on page 50. The difference is that one coordinate is a real number and the other coordinate is an imaginary number. For example, point A in Figure 14.3 has the coordinates 2 and j3. These coordinates combined (2 + j3) give the location of A. Similarly, the other points in Figure 14.3 are:

B = −3 + j2,
C = −1 − j4, and
D = 4 − j3

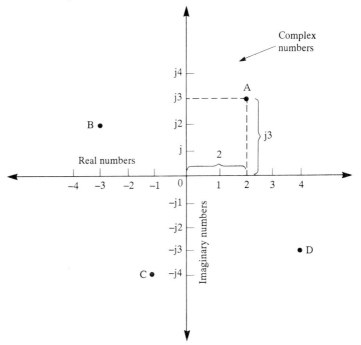

Figure 14.3

Point A is defined by two values, one real, the other imaginary. We say that the point A represents a **complex number**. The complex number consists of a **real part** (2) and an **imaginary part** (3). Note that j does not appear in the imaginary part; it is simply an indicator that the value which follows the j is to be taken in the j direction. This is the way in which j is an operator. We sometimes use the symbols Re and Im to represent the real and imaginary parts of a number. Referring to point B, ReB = −3 and ImB = 2.

Every point in Figure 14.3 represents a complex number. The area of the diagram is known as the **complex number plane**. The real number line is referred to as the **real axis** and the imaginary number line as the **imaginary axis**. Specifying a complex number by its real part and its imaginary part is equivalent to locating a point on a graph by giving its rectangular coordinates. For this reason, a complex number expressed like this is said to be **rectangular form**. Some books refer to this as **standard form**.

Another approach

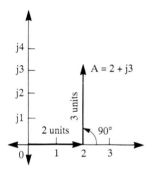

Figure 14.4

One way of getting from the zero to the point A (Figure 14.3) is to travel two units along the real axis, then turn 90° to the left and travel three units upward, in the j direction (parallel to the imaginary axis). In the equation A = 2 + j3, the numbers 2 and 3 indicate the distances to be travelled and the j represents a 90° turn (anticlockwise) between the two stages of the journey. If the diagram is considered as a map of a large field, and a person is standing at the zero point, facing along the real axis in the positive direction, we can issue that person with instructions on how to get to point A. Assuming that the numbers are distances expressed in paces, the instruction 2 + j3 represents: 'Move 2 paces forward; turn 90° left; move 3 paces forward' (Figure 14.4).

Similarly, for point B, the value −3 + j2 means; 'Move 3 paces backward; turn 90° left; move 2 paces forward.' If j is negative, it means turn 90° right so, for point D, 4 − j3 means: 'Move 4 paces forward; turn 90° right; move 3 paces forward.' An alternative interpretation takes −j3 to be j(−3), and the corresponding instruction is: 'Move 4 paces forward; turn 90° *left*; move 3 paces *backward*.' This also brings the person to D.

If j means 'turn 90° left', then j^2 means 'turn 90° left, turn 90° left again'. In other words 'turn 180° left' (Figure 14.5). The person is then facing in the opposite direction, in the *negative* direction along the real axis. Any forward step from zero takes the person to a negative real number. This is consistent with the fact that $j^2 = -1$. In the same way j^3 means turning 3 × 90°, or 270° to the left. The person is now facing in the negative direction of the imaginary axis. This corresponds to the identity, $j^3 \equiv -j$. Note that −j means 'turn 90° right' which also brings the person facing in the negative j direction. Finally j^4 is equivalent to a complete turn, facing in the original direction, since $j^4 \equiv 1$.

The idea of j meaning a 90° turn explains why the real axis and imaginary axis are drawn at right-angles in Figure 14.3. Swinging the real axis about the zero point for a quarter of a turn anticlockwise turns all the real numbers on it into imaginary ones. This is the sense in which j is an operator.

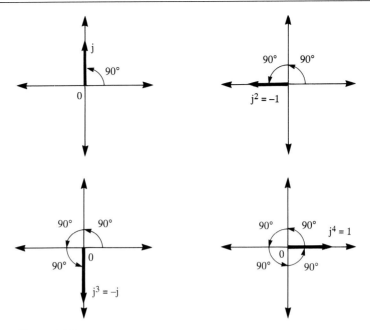

Figure 14.5

The algebra of complex numbers

The essential point to remember when adding or subtracting complex numbers is that the real parts of the numbers and the imaginary parts must be dealt with separately. This is because they represent coordinates in perpendicular directions; an increase or decrease in one coordinate must have no effect on the other.

Example
Add $(4 + j7)$ to $(3 + j2)$

Collect the real parts together and the imaginary parts together:

$\Rightarrow \quad (4 + 3) + (j7 + j2)$

$\quad = 7 + j(7 + 2)$

$\quad = 7 + j9$

This addition is illustrated in Figure 14.6. After a little practice, additions like these can be done mentally.

Examples
$(3 + j5) + (2 - j3) = 5 + j2$

$(2 - j6) + (4 - j) = 6 - j7$

$(5 + j4) - (3 + j3) = 2 - j$

$(2 + j) - (5 - j7) = 7 + j8$

The calculations follow the usual rules. Just as in algebra, terms with j in them must be kept separate.

The same applies to multiplication; we follow the usual algebraic routines.

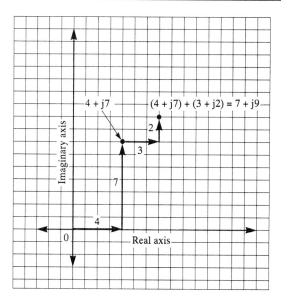

Figure 14.6

Examples

Multiply $(4 + j5)$ by $(3 + j2)$

Set this out as a multiplication:

$$\begin{array}{r} 4 + j5 \\ \times \quad 3 + j2 \\ \hline 12 + j15 \\ j8 + j^2 10 \\ \hline = \ 12 + j23 - 10 \end{array}$$

Add together the imaginary parts. The term which comes from $j5 \times j2$, is now real, with a change of sign, because $j^2 = -1$. Collecting the real numbers together the product is:
$\quad 2 + j23$

More examples

$(3 + j4) \times (2 + j6) = -18 + j26$

$(2 + j5) \times (4 - j2) = 18 + j16$

$(-3 - j5) \times (2 - j4) = -26 + j2$

Conjugate complex numbers

The meaning of conjugate numbers is shown in Figure 14.7. Conjugate numbers differ only in the sign of the imaginary part. In the figure the numbers are $A = 4 + j3$ and $A' = 4 - j3$.

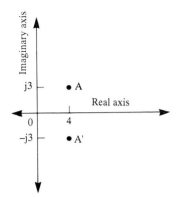

Figure 14.7

Calculating the product of these two numbers:

$$(4 + j3) \times (4 - j3) = 16 + 0j - j^2 9$$

The product is a real number.

In general:

$$(a + jb) \times (a - jb) = a^2 + b^2$$

Division of complex numbers

This appears at first sight to be difficult. For example: divide $(5 + j2)$ by $(3 + j4)$.

Setting this out as a fraction, with numerator and denominator:

$$\frac{(5 + j2)}{(3 + j4)}$$

The problem is how to divide by a complex number. The solution is to convert it into a real number. This is done by using the result obtained above, when a complex number is multiplied by its conjugate. The value of the fraction is unaltered if we multiply both numerator and denominator by the conjugate of the denominator. The conjugate of the denominator is $(3 - j4)$, so the fraction becomes:

$$\frac{(5 + j2) \times (3 - j4)}{(3 + j4) \times (3 - j4)}$$

Multiplying out the denominator gives: $23 + -j14$

Multiplying out the denominator gives: $3^2 + 4^2 = 25$

The value of the fraction is: $\dfrac{23 + -j14}{25} = \dfrac{23}{25} - \dfrac{j14}{25}$

$$= 0.92 + j0.56$$

Another example: $\dfrac{(5 + j3)}{(2 + j)} = \dfrac{(5 + j3) \times (2 - j)}{(2 + j) \times (2 - j)}$

$$= \frac{13 + j}{5} = 2.6 + j0.2$$

> **Test yourself 14.2**
>
> Simplify the following expressions.
> 1. $(2 + j3) + (4 + j2)$
> 2. $(2 - j3) + (3 - j)$
> 3. $(4 - j5) - (4 + j2)$
> 4. $(3 + j7) + (4 - j6)$
> 5. $(a - jb) + 2(b - j4)$
> 6. $(\theta + j\pi) - (3\theta + j2\pi)$
> 7. $(3 + j) \times (2 + j)$
> 8. $(4 + j2) \times (3 - j)$
> 9. $(7 - j3) \times (3 + j2)$
> 10. $(3 + j6)^2$
> 11. $\dfrac{(2 + j)}{(3 + j)}$
> 12. $\dfrac{(4 + j3)}{(2 + j4)}$
> 13. $\dfrac{(1 + j3)}{(1 - j3)}$
> 14. $\dfrac{(3 + j2)}{(2 + j4)}$

Complex numbers and quadratic equations

We now have a way of finding the solution of a quadratic equation when the discriminant (page 40) is less than zero. Take this example:

$$3x^2 - 18x + 39 = 0$$

The coefficients are: $a = 3$, $b = -18$ and $c = 39$

The discriminant is $\sqrt{b^2 - 4ac} = \sqrt{324 - 468}$
$$= \sqrt{-144} = 12\sqrt{-1}$$
$$= j12$$

The discriminant is an imaginary number so the equation has complex roots. When inserted in the quadratic formula, we have:

$$x = \frac{-b \pm \sqrt{b^2 - 4ac}}{2a}$$
$$= \frac{18 \pm j12}{6} = 3 \pm j2$$

Solutions:
$$x = 3 + j2 \quad \text{or} \quad x = 3 - j2$$

Other quadratic equations with negative discriminants are solved in same way.

Complex numbers and vectors

The addition operation illustrated in Figure 14.6 has the features of addition of vectors (page 30). In Figure 14.8 the same addition is re-drawn as a vector addition. A diagram like this, which shows vectors drawn in the complex number plane, is called an **Argand diagram**.

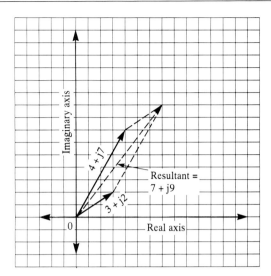

Figure 14.8

One particular kind of vector, used in describing alternating currents and voltages in circuits, is the **phasor** (page 99). The mathematics of imaginary and complex numbers is a useful tool for dealing with these phasors.

Phasors are usually specified by length and by phase angle, for we most often know the amplitude and the phase angle of the sine wave they represent. It is therefore more convenient if the complex numbers that are to represent the phasors are expressed in **polar form**, as a magnitude and an angle.

Polar form

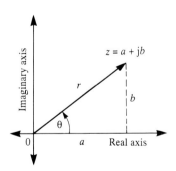

Figure 14.9

Figure 14.9 shows a complex number at point z in the complex number plane. The real part of z is a, and the imaginary part of z is b:

$$z = a + jb \tag{1}$$

This is the way of expressing a complex number in rectangular form. The figure also shows a vector (or phasor) drawn from the origin to point z. The length of the vector is r and its phase angle is θ. From the definition of the trig ratios (page x):

$$a = r \cos \theta \tag{2}$$

$$\text{and} \quad b = r \sin \theta \tag{3}$$

Substituting these values for a and b in equation (1):

$$z = r (\cos \theta + j \sin \theta)$$

This is the **polar form** of the complex number.

Converting from rectangular to polar form

Given a and b, the real and imaginary parts of z, the conversion is exactly the same as on page 100. The fact that b is the imaginary part of z (though not an imaginary number itself) makes no difference to the geometry.

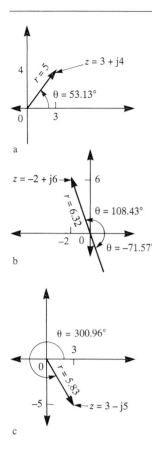

Figure 14.10

Examples

Convert the complex number $z = 3 + j4$ into polar form (Figure 14.10a).

$$r = \sqrt{3^2 + 4^2} = \sqrt{9 + 16} = \sqrt{25} = 5$$

$$\theta = \tan^{-1} 4/3 = 53.13° \text{ (2 dp, by pocket calculator)}$$

In polar form, $z = 5 (\cos 53.13° + j \sin 53.13°)$

Convert $z = -2 + j6$ into polar form (Figure 14.10b).

$$r = \sqrt{2^2 + 6^2} = \sqrt{4 + 36} = \sqrt{40} = 6.32 \text{ (2 dp)}$$

$$\theta = \tan^{-1} 6/-2 = -71.57° \text{ (2 dp, by pocket calculator)}$$

A calculator does not necessarily show the correct result. We might take $-71.57°$ to mean that the point is in the fourth quadrant. But Figure 14.10b shows that the number is in the second quadrant. *Both* the angles shown have a tangent of 3. The calculator does not tell us which angle we require. This is why it is important to draw a sketch (*not* necessarily a scale diagram) when converting to polar form. In this example, the sketch shows that the required value of θ is $180° - 71.57° = 108.43°$. This is the angle measured from the positive direction of the real number line. Using the appropriate angle:

$$z = 6.32 (\cos 108.43° + j \sin 108.43°)$$

Convert $z = 3 - j5$ into polar form (Figure 14.10c)

$$r = \sqrt{3^2 + 5^2} = \sqrt{9 + 25} = \sqrt{34} = 5.83 \text{ (2 dp)}$$

$$\theta = \tan^{-1} -5/3 = -59.04° \text{ (2 dp)}$$

This time the sketch shows that the point is in the fourth quadrant. The value of θ should be $360° - 59.04° = 300.96°$.

In polar form $z = 5.83 (\cos 300.96° + j \sin 300.96°)$.

All complex numbers expressed in polar form have the same format:

$$z = r (\cos \theta + j \sin \theta)$$

It is only the r and θ which vary, so there is no need to repeat the 'cos' and 'j sin' every time we write out a number. The usual convention is to express the number using the symbol shown on page 96. For example, the number $5.83(\cos 300.96° + j \sin 300.96°)$ is written $5.83\underline{/300.96°}$.

Converting from polar to rectangular form

This is simply a matter of applying the equations on page 101.

Examples

Convert $z = 5.39\underline{/21.80°}$ to rectangular form.

$$a = r \cos \theta = 5.39 \times 0.928 = 5$$
$$b = r \sin \theta = 5.39 \times 0.371 = 2$$

The answers are 5 and 2 to two decimal places, so we can take them to be integers within the range of accuracy of the calculation. The rectangular form is:

$z = 5 + j2$

Convert $z = 7.62 \underline{/336.80°}$ to rectangular form.

$a = 7.62 \times 0.92 = 7$

$b = 7.62 \times -0.39 = -3$

In rectangular form, $z = 7 - j3$.

Test yourself 14.3

Solve these quadratic equations.

1 $x^2 - 2x + 10 = 0$ **2** $2x^2 - 10x + 17 = 0$

Convert from rectangular form to polar form (2 dp).

3 $5 + j2$ **4** $3 - j7$
5 $-2 + j3$ **6** $-4 - j5$
7 4 **8** $3 + j3$
9 $j7$ **10** $-6 - j$

Convert from polar form to rectangular form (round to integers).

11 $5 \underline{/36.87°}$ **12** $12 \underline{/4.76°}$
13 $7.07 \underline{/225°}$ **14** $3.61 \underline{/326.31°}$
15 $2.24 \underline{/206.57°}$ **16** $7.62 \underline{/113.2°}$

Negative angles

When a complex number is in the third or fourth quadrant it is often more convenient to measure the angle in the clockwise (negative) direction. Figure 14.11 shows the number $z = 6 - j5$. In polar form this is $7.8 \underline{/-39.8°}$. Written out in full:

$z = 7.8(\cos -39.8° + j \sin -39.8°)$ \hfill (4)

But, as can be seen at the left of Figure 5.24:

$\cos -\theta = \cos \theta$

$\sin -\theta = -\sin -\theta$

Substituting these in equation (4):

$z = 7.8 (\cos 39.8° - j \sin 39.8°)$

This is the same as the usual rectangular form, except that the plus sign is replaced by a minus sign. Therefore, whenever there is a minus sign in the rectangular form, it means that the angle is in the third or fourth quadrant.

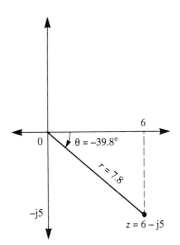

Figure 14.11

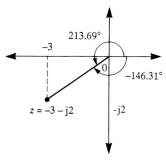

Figure 14.12

Example

$z = -3 - j2 = 3.61 \underline{/213.69°}$

Written out in full:

$z = 3.61(\cos 213.69° + j \sin 213.69°)$

But 216.69° is the same as −146.31° (Figure 14.12), so we can write:

$z = 3.61(\cos 146.31 - j \sin 146.31)$

Both forms define the same point on the number plane.

For negative angles we can use the alternative short form for writing polar coordinates (page 96).

$3.61 \underline{/213.69°} \equiv 3.61 \underline{/-146.31°} = 3.61 \overline{\vert 146.31°}$

Adding complex numbers in polar form

We are sometimes given a pair of vectors, expressed in polar form, and asked to find their sum or **resultant vector**. The way to tackle this is to convert them both to rectangular form, sum them as on page 30, and convert the sum back to polar form.

Example

Sum the vectors $\mathbf{z}_1 = 8.06 \underline{/60.26°}$ and $\mathbf{z}_2 = 3.61 \underline{/33.69°}$

Converting $\mathbf{z}_1$ to rectangular form: $\mathbf{z}_1 = 4 + j7$

Converting $\mathbf{z}_2$ to rectangular form: $\mathbf{z}_2 = 3 + j2$

Summing these: $\mathbf{z}_1 + \mathbf{z}_2 = 7 + j9$

Converting the sum to polar form: $\mathbf{z}_1 + \mathbf{z}_2 = 11.40 \underline{/52.13°}$

This is the summation represented in Figure 14.8.

Multiplying in polar form

Multiplying two complex numbers together is made easier if they are in polar form. It is very much easier than the explanation which follows. If you are mainly interested in the *how* rather than the *why*, skip to the result at the end.

Suppose that we have two complex numbers z_1 and z_2. We want to find their product. In polar form the two numbers are:

$z_1 = r_1 \underline{/\theta_1} = r_1 (\cos \theta_1 + j \sin \theta_1)$

$z_2 = r_2 \underline{/\theta_2} = r_2 (\cos \theta_2 + j \sin \theta_2)$

Their product is:

$z_1 z_2 = r_1 r_2 (\cos \theta_1 + j \sin \theta_1)(\cos \theta_2 + j \sin \theta_2)$

$= r_1 r_2 (\cos \theta_1 \cos \theta_2 + j \sin \theta_1 \cos \theta_2 + j \cos \theta_1 \sin \theta_2 + j^2 \sin \theta_1 \sin \theta_2)$

Rearranging terms, and substituting −1 for j_2, we obtain

$z_1 z_2 = r_1 r_2 [(\cos \theta_1 \cos \theta_2 - \sin \theta_1 \sin \theta_2) + j (\sin \theta_1 \cos \theta_2 + \cos \theta_1 \sin \theta_2)]$

This rather lengthy equation is simplified by making use of two trig identities:

$$\cos\theta_1 \cos\theta_2 - \sin\theta_1 \sin\theta_2 \equiv \cos(\theta_1 + \theta_2)$$
$$\sin\theta_1 \cos\theta_2 + \cos\theta_1 \sin\theta_2 \equiv \sin(\theta_1 + \theta_2)$$

Substituting these in the equation gives:

$$z_1 z_2 = r_1 r_2 \left[\cos(\theta_1 + \theta_2) + j \sin(\theta_1 + \theta_2 0]\right.$$

Or, putting both sides of the equation into the short form:

$$z_1 z_2 = r_1 \underline{/\theta_1} \times r_2 \underline{/\theta_2}$$
$$= r_1 r_2 \underline{/(\theta_1 + \theta_2)}$$

The result is expressed as an easy rule:

To multiply two complex numbers in polar form, multiply the *r*'s and add the θ's.

Examples

$2\underline{/50°} \times 7\underline{/30°} = 14\underline{/80°}$.

$1.5\underline{/100°} \times 6\underline{/-35°} = 9\underline{/65°}$

This rule is extended to multiplication of three or more numbers. Just multiply all the *r*'s and add all the θ's. For example:

$3\underline{/80°} \times 2\underline{/30°} \times 5\underline{/120°} = 30\underline{/230°}$

Dividing in polar form

A similar line of argument to that given above leads to the rule for division:

To divide two number in complex form, divide the *r*'s and subtract the θ's.

Examples

$$\frac{6\underline{/75°}}{3\underline{/25°}} = 2\underline{/50°}$$

$$\frac{5\underline{/30°}}{2\underline{/50°}} = 2.5\underline{/-20°}$$

Powers of complex numbers

It follows from the multiplication rule that to square a complex number we square the *r* and double the θ:

$$(r\underline{/\theta})^2 = r\underline{/\theta} \times r\underline{/\theta} = r^2\underline{/2\theta}$$

For example: $(6\underline{/55°})^2 = 36\underline{/110°}$

Similarly for higher powers:

$$(r\underline{/\theta})^3 = r^3\underline{/3\theta}$$
$$(r\underline{/\theta})^4 = r^4\underline{/4\theta}$$

And for the nth power:

$$(r\underline{/\theta})^n = r^n\underline{/n\theta}$$

This result is known as **De Moivre's theorem**. The theorem also applies to fractional powers, when n is between 0 and 1.

Example
Finding the square root of a complex number.

$$(9\underline{/80°})^{1/2} = 9^{1/2}\underline{/(½ \times 80)} = 3\underline{/40°}$$

This apparently straightforward answer needs to be looked at more closely.

Roots of complex numbers

In the example above we found the square root of $9\underline{/80°}$. Figure 14.13a shows this number and its square root ($3\underline{/40°}$) in an Argand diagram. The number is represented as a vector at an angle of 80°. If we let the vector swing round a whole revolution until it points in the same direction as before, the angle of the vector is 80° + 360° = 440° (Figure 14.13b). The vector also represents the number $9\underline{/440°}$. The square root of *this* number is $3\underline{/220°}$, and this is shown in the figure.

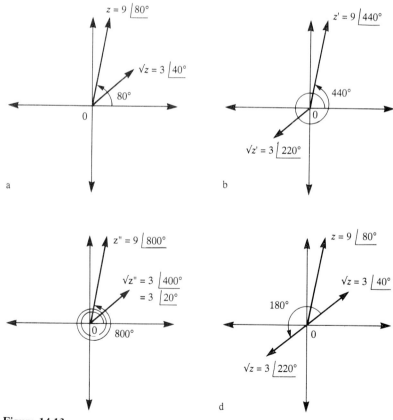

Figure 14.13

Now let the vector make another revolution, so that it represents the number $9\underline{/800°}$ (Figure 14.13c). The square root of this is $3\underline{/400°}$. The angle 400° of the square root gives it the same direction as the first square root, $3\underline{/40°}$. If we continue to add 360° to the number, we find that the square root increases by 180° each time. This is to be expected, since every increase in the angle of the *number* produces an increase of 360°/2 in the angle of the *square root*.

Figure 14.13d illustrates the conclusions from this discussion, that the complex number has *two* square roots, situated 180° apart. This result applies to the square roots of all complex numbers.

In Figure 14.14 we have shown the cube roots of another complex number, $27\underline{/45°}$. Applying the rule, the cube root of 27 is 3 and one-third of 60° is 15°, so the cube root is $3\underline{/15°}$. But we can continue to turn the vector 360° at a time and generate more cube roots. Each turn of 360° increases the angle of the cube root by 360°/3, that is, by 120°. So we obtain *three* cube roots, spaced 120° apart.

Similarly a complex number has four fourth-roots spaced 90° apart, five fifth-roots spaced 72° apart, and so on. In general, a complex number has *n* *n*th-roots, spaced 360°/*n* apart.

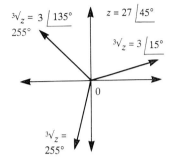

Figure 14.14

Exponential form

As well as the rectangular and polar forms, there is yet another form in which complex numbers are sometimes expressed. This is the exponential form. To understand this we need to know something about **series**. A series is the sum of a number of terms of a **sequence** of numbers (real or imaginary). The numbers of the sequence are calculated according to a set of rules, which may differ from one series to another.

Examples
In the sequence 1, 2, 3, 4, 5, ... the numbers increase by 1 each time. Given the first few numbers, we can easily discover the rule and continue to write out the sequence indefinitely. The sum of the first five terms of the sequence is:

$$S = 1 + 2 + 3 + 4 + 5 = 15$$

As the number of terms increases, the series S increases more and more rapidly. With an infinite number of terms, S is infinitely large. We say the series is **divergent**.

In the sequence 1, 0.5, 0.25, 0.125, 0.0625 ... the successive numbers are obtained by dividing the previous number by 2. The sum of the first five terms of this sequence is:

$$S = 1 + 0.5 + 0.25 + 0.125 + 0.0625 = 1.9375$$

As the number of terms increases, S increases less and less rapidly. It can be shown that the value of the series approaches a limit as the number of terms becomes infinite. The limit of S is 2. This series is said to be **convergent**.

An interesting convergent series is this one:

$$S = 1 + x + \frac{x^2}{2!} + \frac{x^3}{3!} + \frac{x^4}{4!} + \frac{x^5}{5!} + \frac{x^6}{6!} + \ldots$$

The rule for generating the sequence is a little more complicated here. Begin with 1, and multiply by x to get from one term to the next. At the same time

> **Factorial numbers**
>
> Factorial *n* consists of *n* multiplied by $(n-1)$, multiplied by $(n-2)$ and so on down to 1. A factorial number is indicated by an exclamation mark (!) following the number. Only positive integers can be made into factorials.
>
> **Examples**
>
> $3! = 3 \times 2 \times 1 = 6$
>
> $7! = 7 \times 6 \times 5 \times 4 \times 3 \times 2 \times 1 = 5040$
>
> Factorials of quite small numbers are very large. For example $12! = 479\,001\,600$.
>
> *Two special examples*
>
> $1! = 1$, as might be expected.
>
> $0! = 1$, which is unexpected.
>
> This can be explained by working out a sequence of factorials in reverse. Starting with $5!$, we obtain the next earlier term in the sequence by dividing it by 5:
>
> $$\frac{5!}{5} = \frac{5 \times 4 \times 3 \times 2 \times 1}{5} = 4 \times 3 \times 2 \times 1 = 4!$$
>
> Similarly:
>
> $$\frac{4!}{4} = 3!$$
>
> $$\frac{3!}{3} = 2!$$
>
> $$\frac{2!}{2} = 1!$$
>
> and
>
> $$\frac{1!}{1} = 0!$$
>
> But $1! = 1$, so $0! = \dfrac{1}{1} = 1$

divide by a factorial number, increasing it by 1 each time. Although factorials do not appear in the first two terms in the equation above, those terms *are* divided both 0! and 1! respectively, but this is not shown because both 0! and 1! are equal to 1.

The value of the series depends on the value taken for *x*, which must be the same for every term. Suppose that $x = 1$. The value of the first seven terms works out to be:

$$S = 1 + 1 + \frac{1}{2} + \frac{1}{6} + \frac{1}{24} + \frac{1}{120} + \frac{1}{720} = 2.718 \text{ (3 dp)}$$

This result has a familiar appearance (see page 5). Try the series with $x = 2$, summing the first nine terms, as this series converges rather slowly:

$$S = 1 + 2 + \frac{4}{2} + \frac{8}{6} + \frac{16}{24} + \frac{32}{120} + \frac{64}{720} + \frac{128}{5040} + \frac{256}{40320} = 7.387 \ (3\text{ dp})$$

A little research shows that this number is fairly close to e^2, which is 7.389. As we increase the number of terms, the series converges exactly on the value of e^2. Further investigations or a formal proof show that, for any value of x, the value of the series equals e^x. We call it the **exponential series**. Here is a way of calculating e^x to any required degree of precision.

There are two other important series of similar form but differing slightly in the way the terms are generated. These converge on trig functions:

$$\cos x = 1 - \frac{x^2}{2!} + \frac{x^4}{4!} - \frac{x^6}{6!} + \frac{x^8}{8!} - \ldots$$

This has the alternate terms of the exponential series, beginning with the first. The signs are alternately + and −.

$$\sin x = x - \frac{x^3}{3!} + \frac{x^5}{5!} - \frac{x^7}{7!} + \frac{x^9}{9!} - \ldots$$

This has the alternate terms of the exponential series, beginning with the second. The signs are alternately + and −. These series give the values for $\cos x$ and $\sin x$, provided that the angles are expressed in radians.

Above, we substituted values 1 and 2 for x in the exponential series, to obtain values for e and e^2. See what happens if we substitute the imaginary number $j\theta$ to obtain a value for $e^{j\theta}$:

$$e^{j\theta} = 1 + j\theta + \frac{(j\theta)^2}{2!} + \frac{(j\theta)^3}{3!} + \frac{(j\theta)^4}{4!} + \frac{(j\theta)^5}{5!} \ldots$$

$$= 1 + j\theta + \frac{j^2\theta^2}{2!} + \frac{j^3\theta^3}{3!} + \frac{j^4\theta^4}{4!} + \frac{j^5\theta^5}{5!} \ldots$$

Simplifying the powers of j:

$$= 1 + j\theta - \frac{\theta^2}{2!} - \frac{j\theta^3}{3!} + \frac{\theta^4}{4!} + \frac{j\theta^5}{5!} \ldots$$

Collecting even powers of θ and odd powers of θ, also taking out j as a factor from the odd-powered terms:

$$= (1 - \frac{\theta^2}{2!} + \frac{\theta^4}{4!} - \ldots) + j(\theta - \frac{\theta^3}{3!} + \frac{\theta^5}{5!} - \ldots)$$

The expression in the first bracket is the series for $\cos \theta$; the expression in the second bracket is the series for $\sin \theta$, so we obtain the result:

$e^{j\theta} = \cos \theta + j \sin \theta$

The right side of the equation is recognizable as the polar form of a complex number. Thus, if we have a number in polar form, it is converted to exponential form by using the equation:

$$r(\cos\theta + j\sin\theta) = re^{j\theta}$$

For example

$$z = 4\underline{/3}\text{ rad} = 4(\cos 3 + j\sin 3) = 4e^{j3}$$

Remember that in exponential form the angle *must* be in radians. For example,

$$z = 5\underline{/125°} = 5\underline{/2.18}\text{ rad} = 5e^{j2.18}$$

By reasoning similar to that above we obtain a result for negative angles. If:

$$z = r(\cos\theta - j\sin\theta)$$

Then:

$$z = re^{-j\theta}$$

Uses for the exponential form

One way in which the exponential form is often used is when making a Laplace transform (page 201). Putting a number as a power of e makes it suitable for the transformation, since the transform of e^{at} is

$$\frac{1}{s-a}$$

A complex number $re^{j\theta t}$ is transformed to

$$\frac{r}{s-j\theta}$$

Putting a complex number into exponential form also makes it possible to obtain the log of the number. Given that:

$$z = r(\cos\theta + j\sin\theta) = re^{j\theta}$$

Taking natural logs (page 26):

$$\ln z = \ln r + j\theta$$

Similarly, for negative angles:

$$z = re^{-j\theta}$$

$$\Rightarrow \quad \ln z = \ln r - j\theta$$

Test yourself 14.4

Simplify, giving answers in polar form to 2 dp.

1. $2\underline{/45°} + 2\underline{/135°}$
2. $2\underline{/30°} + 3\underline{/60°}$
3. $4\underline{/25°} + 4\underline{/335°}$
4. $2.5\underline{/200°} + 1.5\underline{/315°}$
5. $3\underline{/20°} \times 4\underline{/30°}$
6. $7\underline{/180°} \times 5\underline{/60°}$
7. $2\underline{/10°} \times 3\underline{/-5°}$
8. $10\underline{/-100°} \times 7\underline{/-25°}$
9. $\dfrac{6\underline{/50°}}{2\underline{/20°}}$
10. $\dfrac{15\underline{/30°}}{2\underline{/70°}}$
11. $\dfrac{3\underline{/20°} \times 12\underline{/45°}}{9\underline{/30°}}$
12. $\dfrac{4\underline{/150°} \times 10\underline{/300°}}{5\underline{/200°}}$
13. $(4\underline{/30°})^2$
14. $(3\underline{/100°})^4$
15. $\sqrt[3]{64\underline{/150°}}$
16. $\sqrt[4]{1296\underline{/200°}}$

Convert to exponential form.

17. $7\underline{/2}$ rad
18. $3.2\underline{/22.5}$ rad
19. $4\underline{/180°}$
20. $2.7\underline{/225°}$

Impedances in equations

The impedance of a circuit or part of a circuit is due to one or more of:

- resistance, R
- inductive reactance, X_L
- capacitative reactance, X_C

All three are expressed in ohms. The total impedance Z of a circuit is calculated by considering the various resistances and reactances to be in series or in parallel, according to the way they are connected. When an alternating voltage is applied to the circuit, the different kinds of circuit element behave differently:

Element	Phase angle* of voltage across element in series circuit	Phase angle* of current through element in parallel circuit	Effect of frequency on impedance
Resistance	0°	0°	R is constant
Inductance	+90° (leads)	−90° (lags)	$X_L = \omega L$
Capacitance	−90° (lags)	+90° (leads)	$X_c = \dfrac{1}{\omega C}$

* Relative to applied voltage.

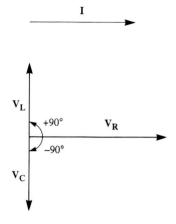

Figure 14.15

Values in the fourth column tell us the *magnitude* of the impedance, from which we can calculate the voltage across the element in a series circuit. But these values do not take account of *phase angle*. In other words, we can calculate the *length* of the voltage phasor, but not the *direction* in which it points. The direction is given in the second column of the table. If we are interested in currents in a parallel circuit, the direction is given in the third column. To be able to handle phasors mathematically, as opposed to drawing a scale diagram and solving problems graphically, we need to be able to represent both magnitude *and* direction in numeric form. Since the j in a complex number represents a rotation of 90°, the complex number system is ideal for this purpose. Using this system, we are able to take the voltage phasors across three types of impedance (Figure 14.15) and express them as complex numbers in polar form:

Impedance due to	Voltage phasor in series circuit	Comment
Resistance	$\mathbf{V_R} = \mathbf{I}\,(R\underline{/0°})$	In phase
Inductance	$\mathbf{V_L} = \mathbf{I}\,(\omega L\underline{/90°})$	Reactance leads
Capacitance	$\mathbf{V_c} = \mathbf{I}\left(\dfrac{-1}{\omega C}\underline{/90°}\right)$	Reactance lags

The equations incorporate the information of the third *and* fourth columns of the previous table. In the equations, the expressions in brackets represent the impedance vectors, $\mathbf{Z_R}$, $\mathbf{Z_L}$ and $\mathbf{Z_C}$ respectively.

Figure 14.16a shows a conventional circuit diagram, and Figure 14.16b shows the components labelled with impedance vectors. These are calculated from the equations in the second column of the table above. If, as here, we know the frequency of the applied voltage, we can incorporate the value of ω into the impedance vectors, as has been done in Figure 14.16c. Using these values and applying the rules for adding, subtracting, multiplying, or dividing complex numbers, we can calculate the vectors for impedances, voltages and currents, in any part of the circuit.

For example, suppose the voltage applied to the circuit of Figure 14.16 has a frequency of 1 kHz. The components have the impedances shown in Figure 14.16c, calculated by substituting $\omega = 2\pi \times 1000 = 6283$ in the values shown in Figure 14.16b. The voltage phasor **V** equals $10\underline{//0°}$, indicating that its magnitude is 10 V. The 0° shows that this is taken as the reference, to which the phase angles of other phasors refer. Now we calculate the current phasors:

$$\mathbf{I_1} = \frac{\mathbf{V}}{100} = \frac{10\underline{/0°}}{100} = 0.1\underline{/0°}$$

$$\mathbf{I_2} = \frac{\mathbf{V}}{j126} = \frac{10\underline{/0°}}{126\underline{/90°}} = 0.0794\underline{/-90°}$$

$$\mathbf{I_3} = \frac{\mathbf{V}}{-j159} = \frac{10\underline{/0°}}{159\underline{/-90°}} = 0.0629\underline{/90°}$$

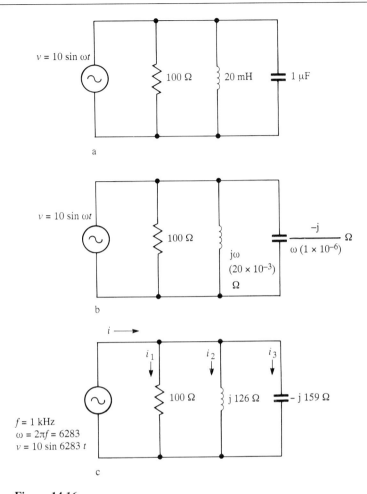

Figure 14.16

Summing these currents to obtain the supply current, after converting them to rectangular form:

$$\mathbf{I} = (0.1 + j0) + (0 - j0.0794) + (0 + j0.0629)$$
$$= 0.1 - j0.0165$$

Converting to polar form:

$$\mathbf{I} = 0.1 \underline{/-9.37°}$$

Figure 14.17 shows this calculation as a phasor diagram. **I** is the resultant of $\mathbf{I_1}$ and $(\mathbf{I_2} - \mathbf{I_3})$. Figure 14.17 is drawn as a scale diagram to confirm that the calculations above give a correct result. Although it often helps to draw a *sketch*, a scale diagram is not needed. Everything can be *calculated* using complex numbers.

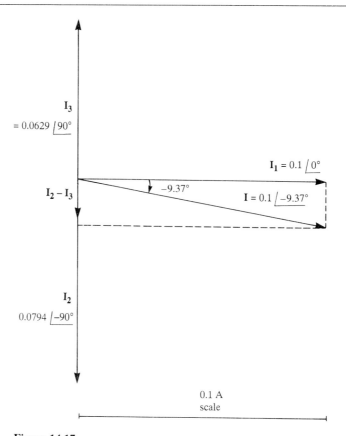

Figure 14.17

Imaginary numbers solve real problems

Here are some examples of electronics calculations which are most easily solved by using complex numbers. Most results are worked to two decimal places.

1 Alternating supply voltage

Problem: Two sections of a circuit are connected in series and an alternating voltage is applied across them. The voltages across the two sections are $v_1 = 40 \sin \omega t$ V and $v_2 = 25 \sin(\omega t - 40°)$ V. What is the rms value of the applied voltage? What is its phase angle with respect to v_1?

Solution: The answer is to be an rms voltage, so we first calculate the rms voltages across the circuit sections by dividing by $\sqrt{2}$.

$$v_{1rms} = \frac{40}{\sqrt{2}} \sin \omega t = 28.28 \sin \omega t$$

$$v_{2rms} = \frac{25}{\sqrt{2}} \sin \omega t = 17.68 \sin(\omega t - 40°)$$

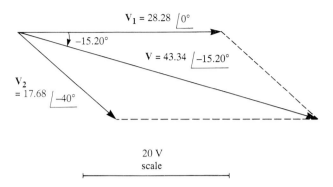

Figure 14.18

Expressing these as phasors in polar form (Figure 14.18):

$\mathbf{V_1} = 28.28\underline{/0°}$ (the angle of this phasor is the reference for the others)

$\mathbf{V_2} = 17.68\underline{/-40°}$ (lags behind $\mathbf{V_1}$)

Since we have to add these phasors, we must convert them to rectangular form:

$\mathbf{V_1} = 28.28 + j0$

$\mathbf{V_2} = 13.54 - j11.36$

Now we can add them:

$\mathbf{V} = \mathbf{V_1} + \mathbf{V_2} = (28.28 + 13.54) + j(0 - 11.36) = 41.82 - j11.36$

To find the rms voltage (the magnitude of the phasor) and the phase angle we convert the sum back to polar form:

$\mathbf{V} = 43.34\underline{/-15.20°}$

The applied rms voltage is 43.34 V and its phase angle is −15.20°.

Comments: Conversion from rectangular form to polar form and the reverse are needed at several stages of this type of calculation. We use bold capital letters as symbols for vectors, including phasors.

Reference pages: 96, 245, 249.

2 Instantaneous voltage

Problem: In the circuit of problem 1, given that the frequency is 50 Hz, calculate the instantaneous value v of the applied voltage 5 ms after the beginning of the cycle of v_1.

Solution: From the equation for the phasor $\mathbf{V}$, we obtain an equation for v:

$v = 43.34\sqrt{2} \sin(\omega t - 15.20)$ V

The $\sqrt{2}$ converts the voltage from its rms value to instantaneous values. We can work in degrees or radians, and choose degrees: If $f = 50$ then $\omega = 360 \times 50 = 18\,000°\ \text{s}^{-1}$ Substituting for ω and t in the equation:

$$v = 43.34\sqrt{2} \sin (18\,000 \times 0.005 - 15.20)$$
$$= 61.29 \sin 74.8 = 61.29 \times 0.965 = 59.15 \quad \text{V}$$

Comments: (1) This problem does not directly involve complex numbers but it illustrates the fact that we may or may not need to know ω and t to solve a given problem. When an alternating current or voltage is quoted in the form $v = A \sin (\omega t + \varphi)$, as in problem 1, we can immediately write out the equation for the phasor without needing to know ω or t:

$$\mathbf{V} = \frac{A}{\sqrt{2}} \underline{/\varphi}$$

This is because angular velocity has no effect on the size of the phasor, and the time since the beginning of the cycle has no effect on the angles between the various phasors.

However, as shown in problem 2, the equation for the resultant phasor $\mathbf{V}$ of problem 1 can be turned back into an equation for instantaneous voltage. Then, given ω and t, we are able to calculate v at any instant.

(2) Make sure ω and φ are *both* in radian measure or *both* in degree measure before adding them.

Reference pages: 66, 99.

3 Alternating supply current

Problem: Two sections of a circuit are wired in parallel, and an alternating voltage is applied across them. The currents are $i_1 = 40 \sin (\omega t + 15°)$ A, $i_2 = 15 \sin (\omega t - 30°)$ A, the phase angles being with reference to the supply voltage. What is the supply current and its phase angle?

Solution: Written in polar form, the current phasors are:

$$\mathbf{I}_1 = 40\underline{/15°} \quad \mathbf{I}_2 = 15\underline{/-30°}$$

Converting to rectangular form, so that they can be added:

$$\mathbf{I}_1 = 38.64 + j10.35 \quad \mathbf{I}_2 = 12.99 - j7.50$$

Adding the phasors:

$$\mathbf{I} = \mathbf{I}_1 + \mathbf{I}_2 = (38.64 + 12.99) - j(10.35 - 7.50)$$
$$= 51.63 + j2.85$$

Converting to polar form:

$$\mathbf{I} = 51.71\underline{/3.16°}$$

The current is 51.71 A, with phase angle 3.16°.

Reference pages: 245, 249.

4 Resonance

Problem: A circuit (Figure 14.19) has two sections in parallel, with the impedances indicated. What value inductor is to be connected where shown, to make the circuit resonate with the applied voltage?

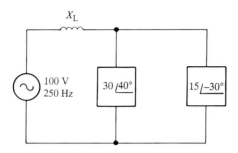

Figure 14.19

Solution: First calculate the impedance $\mathbf{Z_1}$ of the two sections in parallel, using the same formula as for two resistors in parallel (page 77):

$$\mathbf{Z_1} = \frac{30\underline{/40°} \times 15\underline{/-30°}}{30\underline{/40°} + 15\underline{/-30°}}$$

Multiplying out the numerator gives: numerator = $450\underline{/10°}$.

To add the terms in the denominator, first convert to rectangular form and add:

$$\text{denominator} = (22.981 + j19.284) + (12.990 - j7.500)$$
$$= 35.971 + j11.784$$

Then convert to polar form:

$$\text{denominator} = 37.85\underline{/18.14°}$$

Substituting these results:

$$\mathbf{Z_1} = \frac{450\underline{/10°}}{37.85\underline{/18.14°}} = 11.89\underline{/-8.14°} = 11.77 - j1.68$$

Having the numbers in polar form makes the division easy, but we need the result in rectangular form for the next step. For the circuit to resonate, its total impedance must be in phase with the input. This means that the imaginary part of the impedance must be zero (that is, no phase lead or lag). Therefore the reactance $\mathbf{Z_2}$ of the series inductor must be $+j1.68$, to cancel the $-j1.68$ of the parallel sections.

$$\mathbf{Z_2} = j1.68 = jX_L$$

$$\Rightarrow \qquad X_L = 1.68$$

At 250 Hz, $\quad X_L = 2\pi \times 250 \times L = 1571 \; L$

But for resonance X_L must equal 1.68

$\Rightarrow \quad 1571L = 1.68$

$\Rightarrow \quad L = \dfrac{1.68}{1571} = 1.07 \text{ mH}$

A 1.07 mH inductor is required.

Reference pages: 245, 249, 262.

5 Transmission lines

Problem: The impedance of a pair of wires is measured for a signal of given frequency when the opposite ends of the wires are open-circuited ($\mathbf{Z_{OC}}$) and again when they are short-circuited ($\mathbf{Z_{SC}}$), with these results:

$\mathbf{Z_{OC}} = 45\,200\,\underline{/-76°}$

$\mathbf{Z_{SC}} = 0.63\,\underline{/66°}$

Calculate the characteristic impedance, $\mathbf{Z_O}$ given that:

$\mathbf{Z_O} = \sqrt{\mathbf{Z_{OC}} \times \mathbf{Z_{SC}}}$

Solution: Multiplying the two terms:

$\mathbf{Z_O} = \sqrt{45\,200\,\underline{/-76°} \times 0.63\,\underline{/66°}}$

$\phantom{\mathbf{Z_O}} = \sqrt{28\,476\,\underline{/-10°}}$

Taking the square root:

$\mathbf{Z_O} = 169\ \Omega\,\underline{/-5°}$

Reference pages: 252, 254.

Test yourself 14.5

1 Calculate the current **I** in Figure 14.20a, and its phase relationship to the applied voltage. Give the result in both polar and rectangular form to 2 dp. [Hint: **I = V/Z**.]

2 For the circuit of Figure 14.20b, calculate the input impedance and the input current, in polar form. Is the circuit resonating?

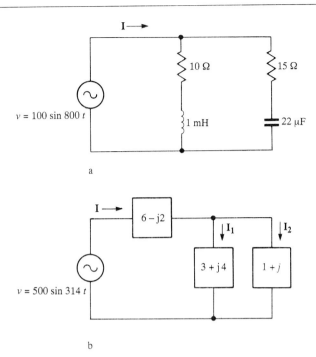

Figure 14.20

15 Analysing waveforms

> In audio and communications applications of electronics it is often necessary to be able to analyse a signal to find out what frequencies are present in it, and in what proportions. The Fourier series is the key to a powerful analytical technique intended for this purpose.
>
> You need to know about functions (page 72), sines and cosines (page 65), integration (Chapter 10, page 138), phasors (page 99) and series (page 255).

The idea behind the Fourier series is that any function may be thought of as the sum of one or more sine functions of different frequencies and different amplitudes. Putting this in terms of audio signals, any periodic waveform is made up of one or more sine waves. But the Fourier series is more general than this, allowing mathematically defined functions such as $y = 3x + 2$ and $y = 2x^3$, to be expressed in terms of sines. Purely random waveforms such as white noise, and the sounds made by many kinds of percussion instrument, cannot be analysed in this way.

When the string of an instrument such as a violin is plucked or bowed, it vibrates and produces a sound. The ends of the string are stationary because they are firmly tightened against the body of the instrument at the bridge and by the finger of the player. But the centre of the string swings widely to and fro (Figure 15.1a). When it is vibrating in this way the string produces its note of lowest pitch. It is the note of lowest frequency, called the **fundamental**, or **first harmonic**, and referred to as f. The points where the string is stationary are known as **nodes** and the point where it is vibrating most widely is known as an **antinode**.

A string is able to vibrate in other ways. In Figure 15.1b we show it vibrating with nodes at its ends and at its centre. There are two antinodes. The distance between the nodes is half that of Figure 15.1a, and the frequency is double, or $2f$. Musicians say that it is an **octave** above the fundamental. It has higher pitch and is called the **second harmonic**.

In Figure 15.1c we see how the string produces its third harmonic, frequency $3f$, vibrating with nodes at its ends and at two evenly-spaced locations along its length. There *must* always be nodes at its ends, which are held against the body of the instrument. In a similar way the string produces fourth, fifth and even higher harmonics. At any given instant, the string is vibrating in all these ways at the same time, producing its fundamental, plus its second, third and several higher harmonics. The amplitude of vibration varies from one harmonic to another with fundamental and lower harmonics usually being the loudest. Certain of the harmonics may make the air in the body of the instrument resonate, making those harmonics louder. The result is a particular mixture of harmonics which we recognize as the distinctive sound of the violin. As the player varies the length of the string by placing a finger

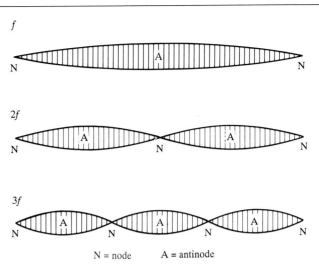

Figure 15.1

at different positions, the fundamental note is changed and also the collection of harmonics which belong to it.

The same principles apply to the production of sound by other musical instruments, and to the production of sound by the human vocal cords. Electronic circuits, too, particularly those of oscillators, amplifiers and filters may vibrate at or resonate at given frequencies and their harmonics. When designing and testing audio and communications circuits, the ability to sort out the response to the component waveforms of an electrical signal is very important and it is here that the Fourier series is so useful.

Sine waves

Sine waves, or **sinusoidal waves**, as they are more correctly called, are described by the basic function:

$$y = A \sin \omega t$$

in which A is the amplitude
ω is the angular velocity (see below) and
t is the independent (time) variable.

The angular velocity in radians per second is directly related to the frequency, f:

$$\omega = 2\pi f$$

so the function is often written in the form:

$$y = A \sin 2\pi f t$$

The period of this function (the number of seconds taken for each cycle) is $1/f$. If we take the period to be 2π, then:

$$f = \frac{1}{2\pi}$$

and the function is simplified to:

$$y = A \sin t$$

All we need to remember is that, using this simplified version, the length of one cycle is 2π seconds.

In Figure 15.2a the fundamental frequency is shown as a graph, for one whole cycle. The equation for this is $y = A_1 \sin t$, using A_1 instead of plain A as we shall be needing several other amplitudes. In Figure 15.2b the second harmonic, having double the frequency, goes through two cycles in the same time. The equation for the second harmonic is:

$$y = A_2 \sin 2t$$

Doubling up the angle before we take the sine means that we fit in *two* cycles instead of one as t increases from 0 to 2π. A_2 does not necessarily have the same value as A_1; the two waves are related in frequency but may differ in amplitude.

Continuing with this, we find that the third harmonic (Figure 15.2c) has the equation:

$$y = A_3 \sin 3t$$

and, in general, for the nth harmonic:

$$y = A_n \sin nt$$

These are the related harmonics which are used to build up multi-frequency periodic functions, such as those representing the sound of a clarinet. A periodic function built up from the waveforms of Figure 15.2a and c is shown at d.

Building the series

A function representing the combination of a fundamental and its harmonics can be written as the sum of the harmonics we have just described:

$$y = A_1 \sin t + A_2 \sin 2t + A_3 \sin 3t \ldots + A_n \sin nt + \ldots$$

Each of the component signals is present in different proportions according to its amplitude. The signals, although running at frequencies which are multiples of the fundamental, are not necessarily in phase with the fundamental or with each other. We allow for this by giving each component a phase angle of its own. Now the series becomes:

$$y = A_1 \sin(t + \varphi_1) + A_2 \sin(2t + \varphi_2) + A_3 \sin(3t + \varphi_3) + \ldots$$
$$+ A_n \sin(nt + \varphi_n) + \ldots$$

There is one further addition to the series, especially appropriate to electronic circuits. Very often an alternating signal does not alternate about 0 V, but has a constant voltage (a dc voltage) added to or subtracted from it. We incorporate this dc voltage in the function, as A_0:

$$y = A_0 + A_1 \sin(t + \varphi_1) + A_2 \sin(2t + \varphi_2) + A_3 \sin(3t + \varphi_3) + \ldots$$
$$+ A_n \sin(nt + \varphi_n) + \ldots$$

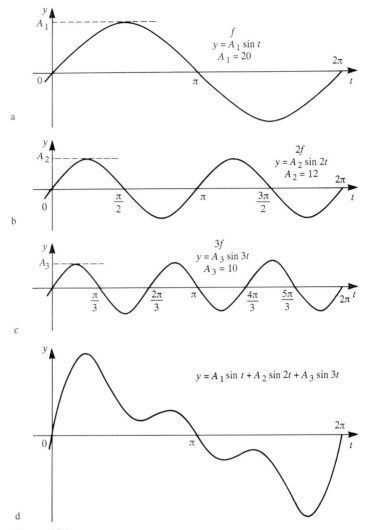

Figure 15.2

This series describes any periodic waveform. It might be wondered how it could possibly describe waves as sharp-cornered, as square waves and triangular waves, but we shall show later that this can be done. For the present, we take this series and convert it into a form that is easier to use.

Converting the sines

With the exception of the first term, the series consists of terms of the form:

$$A_n \sin (nt + \varphi_n)$$

This is expanded by using the trig identity, $\sin (A + B) = \sin A \cos B + \cos A \sin B$, listed on page 68:

$$A_n \sin (nt + \varphi_n) = A_n[\sin nt \cos \varphi_n + \cos nt \sin \varphi_n]$$

Rearranging terms and the order of multiplying (which has no effect of the values):

$$= (A_n \sin \varphi) \cos nt + (A_n \cos \varphi_n) \sin nt$$

The terms in brackets have constant value, because amplitude and phase angle are constant. To simplify the expression, replace these terms with constants, $a_n = A_n \sin \varphi$ and $b_n = A_n \cos \varphi_n$. The terms of the series become:

$$a_n \cos nt + b_n \sin nt$$

Now we are ready to re-write the series in its new form, substituting a new constant $a_0/2$ for A_0, and listing the cosine terms first, followed by the sine terms:

$$y = \tfrac{1}{2}a_0 + a_1 \cos t + a_2 \cos 2t + a_3 \cos 3t + \ldots + a_n \cos nt + \ldots$$
$$+ b_1 \sin t + b_2 \sin 2t + b_3 \sin 3t + b_n \sin nt \ldots$$

This is the form in which the Fourier series is most often written. When analysing a waveform the usual aim is to discover the values of the constant coefficients $a_0, a_1, a_2 \ldots b_1, b_2 \ldots$.

It may seem as if evaluating a Fourier series is a long and tedious matter, but this is usually not so. Often the series converges rapidly so that it reaches a value close enough to its limiting value after only a few terms. Also it may happen that the series does not contain *any* cosine terms (all the *a* coefficients are zero) or any sine terms. If so, the amount of calculation is halved immediately. In this and other ways, to be explained later, the evaluation of a Fourier series is often a relatively short calculation.

Conditions

The Fourier series applies to a function with period 2π, this being the time for exactly one cycle of the fundamental and for two or more complete cycles of the harmonics. In other words, the domain of the function is 2π. We may choose to calculate the function from *any* point during the cycle, provided that we take it up to a time exactly 2π seconds later. Since the fundamental repeats itself every 2π seconds, the choice of starting time makes no difference to the final result. A suitable choice of starting time may make the calculation easier. Usually it is best to cover the period 0 to 2π seconds. Occasionally it is easier if we begin at $-\pi$ and end at $+\pi$. Since the period is taken to be 2π, the frequency of the fundamental is 0.16 Hz. This does not restrict Fourier analysis to signals at 0.16 Hz. In a signal of any frequency the relationship between the fundamental and its harmonics is *relative*. Given any f we know that the harmonics are $2f$, $3f$, $4f$, and so on. The value of f itself is not important in the analysis and so we take the period as 2π to simplify the equations.

Not every function can be represented by a Fourier series. For a function to be analysed, it must conform to a number of conditions, called the **Dirichlet conditions**. As long as these conditions apply over the whole domain of the function, the function can be represented as a Fourier series. The conditions are:

- The function must have a *single* defined value for each value in its domain. An equation such as $x^2 + y^2 = r^2$, which represents a circle, gives *two* values of y for each value of x, so it cannot be represented. Indeed, it is usually not considered to be a function at all.
- The function must not have any infinite discontinuity of range within its domain. For example, $y = \tan x$ does not fulfil this condition as it jumps between $-\infty$ and $+\infty$ when $x = 0$, $x = 2\pi$, $x = 4\pi$, and so on.

  Similarly, $y = \dfrac{2}{x + 3}$ makes a similar jump when $x = -3$.

  But $\dfrac{2}{x - 7}$ has its discontinuity at $x = 7$,

  which is outside the domain 0 to 2π (x is in radians) and fulfils this condition.
- The first and second differentials of the function must be piecewise continuous over the domain.

Most examples taken from electronic circuits conform to the Dirichlet conditions, so this is something that does not normally cause difficulties. If the conditions are all fulfilled, the series is convergent, usually rapidly convergent, and can be evaluated with sufficient precision by taking only the first few terms.

Finding the first term

The first term a_0 is found by integrating the series from 0 to 2π. We will not explain *why*, but will show that it works. We begin by rewriting the series as the first term plus the sum of the cosine and sine terms:

$$y = \tfrac{1}{2}a_0 + \sum_{n=1}^{\infty} (a_n \cos nt + b_n \sin nt)$$

in which n is a positive integer. Integrating both sides of this equation:

$$\int_0^{2\pi} y\, dt = \tfrac{1}{2}\int_0^{2\pi} a_0\, dt + \sum_{n=1}^{\infty} \left(\int_0^{2\pi} a_n \cos nt\, dt + \int_0^{2\pi} nb_n \sin nt \right)$$

But $\cos nt$ and $\sin nt$ integrated from 0 to 2π with respect to t are both zero (see page 154 and the box), so:

$$\int_0^{2\pi} y\, dt = \tfrac{1}{2}[a_0 t]_0^{2\pi} + 0$$

$$= \tfrac{1}{2}[a_0 \cdot 2\pi - a_0 \cdot 0]$$

$$= a_0 \pi$$

$$\Rightarrow a_0 = \frac{1}{\pi} \int_0^{2\pi} y\, dt \qquad (1)$$

This is the equation used in calculating the first term of the series, the first term being $a_0/2$.

> **Definite integrals of sine and cosine functions**
>
> Some integrals reduce to either zero or π when taken over a single cycle, greatly simplifying many of the Fourier calculations. Although the limits used below are 0 to 2π, the integrals have the same value over the limits $-\pi$ to $+\pi$ or any other interval of 2π.
>
> n and m are positive integers, and integration is with respect to the time-variable, t.
>
> Here is a worked example:
>
> $$\int_0^{2\pi} \sin nt \, dt = \left[\frac{-\cos nt}{n} \right]_0^{2\pi} = \frac{1}{n}[-\cos 2n\pi + \cos 0] = \frac{1}{n}[-1 + 1] = 0$$
>
> Similarly:
>
> $$\int_0^{2\pi} \cos nt \, dt = 0$$
>
> $$\int_0^{2\pi} \sin^2 nt \, dt = \pi \qquad \text{given } n \neq 0$$
>
> $$\int_0^{2\pi} \cos^2 nt \, dt = \pi \qquad \text{given } n \neq 0$$
>
> $$\int_0^{2\pi} \sin nt \, \cos mt = 0$$
>
> $$\int_0^{2\pi} \cos nt \, \cos mt = 0 \qquad \text{given } n \neq m$$
> $$\qquad\qquad\qquad\;\; \text{or} = \pi \qquad \text{given } n = m$$
>
> $$\int_0^{2\pi} \sin nt \, \sin mt = 0 \qquad \text{given } n \neq m$$
> $$\qquad\qquad\qquad\;\; \text{or} = \pi \qquad \text{given } n = m$$

Figure 15.3 illustrates this calculation with reference to a sawtooth waveform in which a simple ramp function is repeated with period 2π. The function is:

$y = 3t \qquad 0 < t < 2\pi$

Applying equation (1), and integrating this from 0 to 2π:

$$a_0 = \frac{1}{\pi} \int_0^{2\pi} 3t \, dt = \frac{1}{\pi} \left(\frac{3t^2}{2} \right)_0^{2\pi} = \frac{1}{\pi} \left(\frac{3 \cdot (2\pi)^2}{2} - 0 \right)$$

$$\Rightarrow \qquad\qquad\qquad a_0 = 6\pi$$

From this, we obtain the first term of the series, $a_0/2 = 3\pi$.

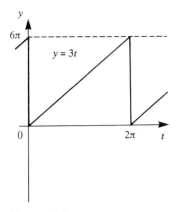

Figure 15.3

Figure 15.3 shows that y increases steadily from 0 to 6π during one cycle. Its mean value is 3π, which is the same as that which we have just calculated for the first term. This illustrates the rule that:

First term = mean value of y over one cycle

This rule often makes calculations very much easier. If it is possible to find the mean value of y by drawing a sketch of the function and applying elementary geometry, we can avoid the integration. Figure 15.4 shows some examples of functions in which a_0, and hence $a_0/2$, can be found by inspecting the sketch. In Figure 14.5a the function has the same waveform as that of Figure 15.3, but the mean value of y is zero. In terms of voltages, the signal of Figure 15.3 can be thought of as the signal of Figure 15.4a superimposed on a dc level of 3π volts. The first term of the Fourier series represents any dc voltage (or steady current or other physical quantity) that may be present.

Figure 15.4b shows the function

$$y = 4 + 2\sin t \quad 0 < t < 2\pi$$

Integrating from 0 to 2π:

$$a_0 = \frac{1}{\pi}\int_0^{2\pi} 4 + 2\sin t \, dt = \frac{1}{\pi}[4t - 2\cos t]_0^{2\pi}$$

But $\cos 2\pi = \cos 0 = 1$:

$$a_0 = \frac{1}{\pi}[(8\pi - 2) - (0\pi - 2)]$$

$$= \frac{1}{\pi}(8\pi)$$

$$= 8$$

From this, the first term in the series is $a_0/2 = 4$. The symmetry of the curve shows that the mean value of y between 0 and 2π is 4. This value could have been obtained *without* any calculation, merely by inspecting the graph, but we calculated it just to illustrate the principle involved.

The piecewise function of Figure 15.4c is a pulsed waveform. Its value is 4 for three-quarters of the time and zero for the remainder. Without further calculation, we can say that $a_0 = 3$ and that the first term is $a_0/2 = 1.5$.

Finding the cosine terms

These are the terms which have $a_1, a_2, \ldots a_n, \ldots$ as their coefficients. The equation for a_n is:

$$a_n = \frac{1}{\pi}\int_0^{2\pi} y \cos nt \, dt$$

As an example, we return to the waveform of Figure 15.3, for which $y = 3t$:

$$a_n = \frac{1}{\pi} \int_0^{2\pi} 3t \cdot \cos nt \, dt$$

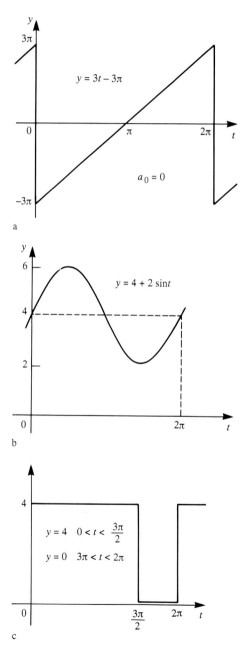

Figure 15.4

This is integrated by parts (page 150):

Let $u = 3t$, then $du/dt = 3$

Let $dv/dt = \cos nt$, then $v = \dfrac{1}{n} \sin nt$

$$a_n = \dfrac{1}{\pi} \left\{ \left[\dfrac{3t}{n} \sin nt \right]_0^{2\pi} - \dfrac{1}{n} \int_0^{2\pi} 3 \sin nt \, dt \right\}$$

But the integral on the right is zero (see box):

$$a_n = \dfrac{1}{\pi n} \{ 6\pi \sin 2n\pi - 6\pi \sin 0 \}$$

The values of $\sin 0$, $\sin 2\pi$, $\sin 4\pi$, $\sin 6\pi$, and so on, are all zero:

$$a_n = 0$$

All the a coefficients are zero and therefore there are no cosine terms in the series for this function.

Multiples of π

If n is an integer, positive or negative and including $n = 0$, then:

$\sin n\pi = 0$

$\cos n\pi = 1$ if n is even

$ = -1$ if n is odd

Finding the sine terms

The routine is similar to that described for cosine terms, except that the equation is:

$$b_n = \dfrac{1}{\pi} \int_0^{2\pi} y \sin nt \, dt$$

Continuing the example of Figure 15.3:

$$b_n = \dfrac{1}{\pi} \int_0^{2\pi} 3t \sin nt \, dt$$

Integrating by parts gives:

$$b_n = \dfrac{1}{\pi} \left\{ \left[-\dfrac{3t}{n} \cos nt \right]_0^{2\pi} + \dfrac{1}{n} \int_0^{2\pi} 3 \cos nt \, dt \right\}$$

The integral on the right equals zero (see box):

$$b_n = \dfrac{1}{\pi n} \{ -6\pi \cos 2n\pi + 0\pi \cos 0 \}$$

The cosine of even multiples of π is 1, so:

$$b_n = \frac{-6}{n}$$

Thus the general form of the sine terms is $\frac{-6}{n} \sin nt$.

The final analysis

We have now established that the Fourier series for $y = 3t$ has the following features:

- $a_{o/2} = 3\pi$
- there are no cosine terms
- the sine terms have the form $\frac{-6}{n} \sin nt$

Using these results, we write out the series for as many terms as we need:

$$y = 3\pi - \frac{6}{1} \sin 1t - \frac{6}{2} \sin 2t - \frac{6}{3} \sin 3t - \frac{6}{4} \sin 4t - \ldots$$

This is more simply written:

$$y = 3\pi - 6 \sin t - 3 \sin 2t - 2 \sin 3t - 1.5 \sin 4t - \ldots$$

Now to put it to the test. Normally we would accept the analysis as correct, assuming that all the calculations have been worked properly. As this is the first example we have worked, we will see that happens when this function is plotted as a graph. In Figure 15.5 (see box, page 282) the graph shows the function calculated as far as the fifth sine term, the fifth harmonic. A short computer program was used to calculate the points. Although the waveforms of the harmonics show up on the graph as slight undulations, the overall shape of the curve is the same as that of Figure 15.3. This analysis is sufficiently precise for many purposes but we can extend the calculation to include more terms if required, and obtain an even better approximation to Figure 15.3. In Figure 15.6 the function is calculated to the twentieth harmonic. There are more ripples but they are much smaller. This is a good demonstration of the way in which a sharply-pointed function such as this sawtooth wave can be analysed into a series of 'curvy' sine waves.

Fourier coefficients

In some applications we do not need to know the equation for the whole series but only the coefficient of one of the harmonics. Ignoring the sign, this gives the amplitude of the harmonic. It is calculated using the techniques explained above. For example, if we need to know the amplitude of the third harmonic of $y = 3t$, we calculate it from

$$b_n = \frac{-6}{n},$$

ignoring the sign. The amplitude of the third harmonic is 6/3 = 2. More important is the amplitude of a harmonic relative to that of the fundamental.

ANALYSING WAVEFORMS

Fourier coefficients

$$a_0 = \frac{1}{\pi}\int_0^{2\pi} y\, dt;\quad \text{First term} = \frac{a_0}{2}$$

cosine terms, $\quad a_n = \dfrac{1}{\pi}\displaystyle\int_0^{2\pi} y\cos nt\, dt$

sine terms, $\quad b_n = \dfrac{1}{\pi}\displaystyle\int_0^{2\pi} y\sin nt\, dt$

Limits of integrals can also be $-\pi$ to π.

Fourier series for $y = 3t$
$y = 3\pi - 6\sin t - 3\sin 2t - 2\sin 3t - 1.5\sin 4t - 1.2\sin 5t$

Figure 15.5

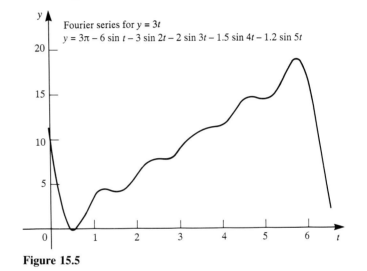

Fourier series for $y = 3t$
$y = 3\pi - 6\sin t - \ldots - 0.3\sin 20t$

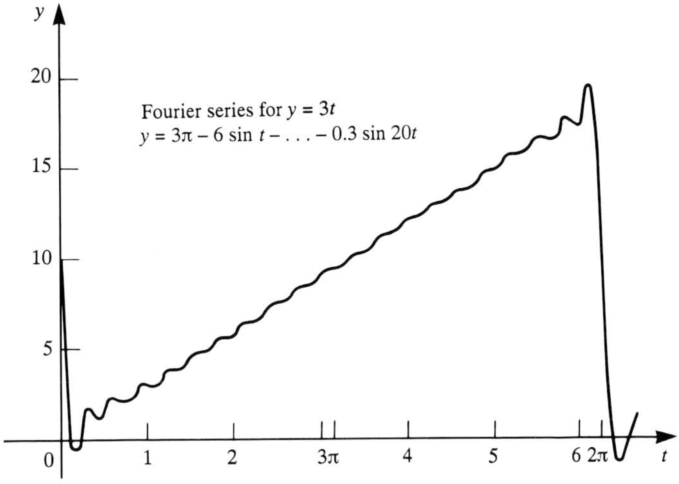

Figure 15.6

This is usually expressed as the **percentage harmonic**. In this example the first harmonic has amplitude 6, so the percentage third harmonic is:

$$\frac{\text{amplitude of third harmonic}}{\text{amplitude of fundamental}} \times 100 = \frac{2}{6} \times 100 = 33.3\% \quad (1\ dp)$$

A graph of the amplitudes of the harmonics is known as the **frequency spectrum** of a function. Figure 15.7 shows the frequency spectrum of the waveform of Figure 15.3. Compared with the spectrums of other functions that we shall look at, this series converges slowly. The coefficient $6/n$ falls more and more slowly as n increases. This accounts for the fact that we need about 20 terms (Figure 15.7) to make it approximate reasonably to the sawtooth waveform.

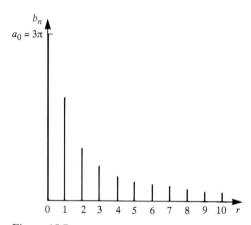

Figure 15.7

In this example the series comprised only the initial (dc) term and sine terms. In series which have both cosine and sine terms it is necessary to sum the corresponding sine and cosine terms for a given harmonic:

$a_n \cos nt + b_n \sin nt$

We cannot simply add a_n to b_n. This is because the two components, being a cosine and a sine respectively, are not in phase with each other. But the cosine of an angle equals the sine of that angle plus 90° (Figure 5.24). Thus the nth harmonic can also be written:

$a_n \sin (nt + 90°) + b_n \sin nt$

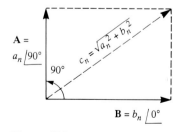

Figure 15.8

In Figure 15.8 these two terms are represented by phasors, with the a_n phasor leading the b_n phasor by 90°. From Pythagoras' theorem, the resultant phasor has magnitude:

$c_n = \sqrt{(a_n^2 + b_n^2)}$

An obvious example

The curve of Figure 15.2b consists only of a dc component plus the fundamental, as can be seen by glancing at its equation:

$$y = 4 + 2 \sin t \quad 0 < t < 2\pi$$

It is hardly necessary to use Fourier analysis on such a trivial function, but we will do so in order to run through the routine with the minimum of distractions and also to show how the definite integrals of the box (page 274) are put to useful effect.

It has already been shown (page 275) that the first term is $a_0/2 = 4$. The formula of a_n gives:

$$a_n = \frac{1}{\pi} \int_0^{2\pi} (4 + 2 \sin t) \cos nt \, dt$$

$$= \frac{1}{\pi} \left(\int_0^{2\pi} 4 \cos nt + \frac{1}{\pi} \int_0^{2\pi} 2 \sin t \cos nt \, dt \right)$$

The equations in the box show that the first integral is zero. The second, in which $m = 1$, is also zero. Thus $a_n = 0$ and there are no cosine terms.

The formula for b_n gives:

$$b_n = \frac{1}{\pi} \int_0^{2\pi} (4 + 2 \sin t) \sin nt \, dt$$

$$= \frac{1}{\pi} \left\{ \int_0^{2\pi} 4 \sin nt \, dt + \frac{1}{\pi} \int_0^{2\pi} 2 \sin t \cos nt \, dt \right\}$$

The first integral is zero. The value of the second integral depends upon the value of n (see box, except that we have n and m interchanged). Here $m = 1$, and if $n \neq m$, then the integral has zero value. However, if $n = 1$ the integral has the value π, and

$$b_1 = 2$$

The series has only one sine term, when $n = 1$. From these calculations we write out the series:

$$y = a_0/2 + b_1 \sin t$$
$$= 4 + 2 \sin t$$

This is an exact solution, and one that we knew already.

A less obvious example

The pulse waveform of Figure 15.4c is one of the more complicated functions and illustrates some more of the paths an analysis may take. This is a piecewise function (page 73):

$$y = 4 \quad 0 < t < \frac{3\pi}{2}$$

$$y = 0 \quad \frac{3\pi}{2} < t < 2\pi$$

It has already been shown that $a_0/2 = 3$.

Integrating for a_n is done separately for each part of the function; the essential point is that the whole period 0 to 2π *must* be covered:

$$a_n = \frac{1}{\pi} \int_0^{3\pi/2} 4 \cos nt \, dt + \frac{1}{\pi} \int_0^{2\pi} 0 \cos nt \, dt$$

In this example, the function has zero value during the later part of the phase, so the second integral is zero, and we evaluate only the first integral. In other examples, where the function has non-zero value at all stages, we evaluate the corresponding integrals and sum the results. Continuing with the first integral:

$$a_n = \frac{4}{\pi} \int_0^{3\pi/2} \cos nt \, dt = \frac{4}{\pi n} [\sin nt]_0^{3\pi/2}$$

Note that the equations in the box on page 274 are of no use in this example, because we are integrating from 0 to $3\pi/2$, not from 0 to 2π.

$$a_n = \frac{4}{\pi n} \left[\sin \frac{3n\pi}{2} - \sin 0 \right]$$

There are cosine terms in the series. Ignoring sin 0, which has zero value, the *a* coefficients are:

$$\frac{4}{\pi n} \sin \frac{3n\pi}{2}$$

Calculating the value of $\sin 3n\pi/2$ for $n = 1, 2, 3, \ldots$ shows that it cycles through four values: $0, -1, 0, 1, \ldots$ This means that there are no odd cosine terms and that the signs of the even terms are alternately negative and positive:

$$\frac{4}{\pi} \left[-\cos t + \frac{\cos 3t}{3} - \frac{\cos 5t}{5} + \frac{\cos 7t}{7} - \ldots \right]$$

A similar calculation shows that:

$$b_n = \frac{4}{\pi n}\left[\cos\frac{3n\pi}{2} - \cos 0\right]$$

in which $\cos 0 = 1$. Calculating b for $n = 1, 2, 3, \ldots$ shows that it cycles through four values: 1, 2, 1, 0, ... Terms are all positive but every fourth term is missing, starting with $n = 4$:

$$\frac{4}{\pi}\left[\sin t + \sin 2t + \frac{\sin 3t}{3} + \frac{\sin 5t}{5} + \frac{\sin 6t}{3}\right]$$

Note the effect of the factor 2 when $n = 2, 6, 10$, etc.

Combining the results of these calculations, the Fourier series is:

$$y = 3 + \frac{4}{\pi}\left[-\cos t + \frac{\cos 3t}{3} - \frac{\cos 5t}{5} + \frac{\cos 7t}{7} - \frac{\cos 9t}{9} + \cdots \right.$$

$$\left. + \sin t + \sin 2t + \frac{\sin 3t}{3} + \frac{\sin 5t}{5} + \frac{\sin 7t}{7} + \frac{\sin 9t}{9} \cdots \right]$$

A plot of this series (Figure 15.9) taken to $n = 12$ shows that it approximates very closely to the original function.

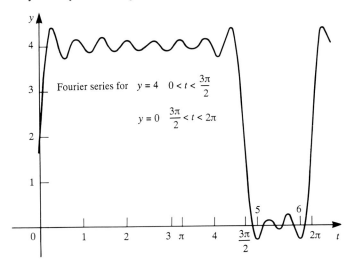

Figure 15.9

Calculating coefficients

The example above is a series in which most of the harmonics have both cosine and sine components. To calculate the amplitude of these harmonics we use the formula given on page 281. For example, the seventh harmonic is:

$$\frac{4}{\pi}\left(\frac{\cos 7t}{7} + \frac{\sin 7t}{7}\right)$$

The coefficients are $a_7 = b_7 = \dfrac{4}{7\pi}$

Applying the formula:

$$c_7 = \sqrt{a_7^2 + b_7^2}$$

$$= \dfrac{4}{\pi}\sqrt{\dfrac{1}{7^2} + \dfrac{1}{7^2}}$$

$$= 0.26$$

The coefficients for the first 13 harmonics are shown in the frequency spectrum of Figure 15.10. After the third harmonic the series converges fairly rapidly. The 4th, 8th, 12th, ... harmonics are absent.

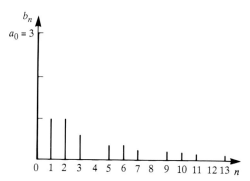

Figure 15.10

> **Test yourself 15.1**
>
> Write out the function for each of the waveforms shown in Figure 15.11. Calculate a_0, a_n and b_n, and write out the first eight terms of the Fourier series for each. Draw the frequency spectrum for the first eight harmonics of each waveform.
>
> [Hint: if you find that the calculations for Figure 15.11d become too lengthy to manage, delay answering this problem until you have read about symmetry.]

Symmetry

The exercises above, illustrate the fact that whole groups of terms are often completely absent from a series. There may be no cosine terms, as in Figure 15.11a, or there may be no sine terms, as in Figure 15.11b. In other series there may be only the terms for which n is odd, as in Figure 15.11d.

It can be shown that waveforms of certain types *always* have the same groups of terms missing from their series. If we know what type the waveform belongs to, we know what groups of terms are absent and so avoid wasting time by trying to calculate them.

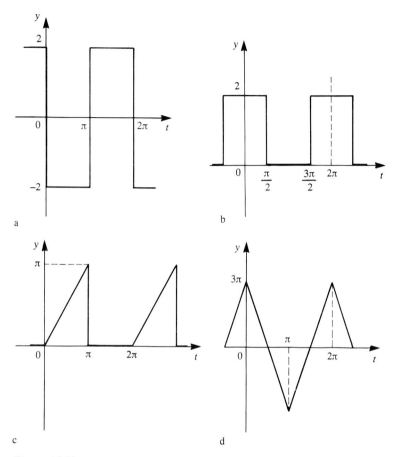

Figure 15.11

The rules for identifying these kinds of waveforms depend on the symmetry of the curve. In one type of symmetry, the waveform of the first half of the cycle is repeated exactly during the second half, but is inverted. One of the simplest examples is the sine curve (Figure 15.12). Other examples are shown in Figures 15.4a, 15.11a and 15.11d. This type of symmetry is known as **half-wave inversion**. One feature is that the mean value of y is zero, so there is no initial term in the series. A further feature of half-wave inversion is that the series has no terms for the even values of n. Figure 15.4b does not show half-wave inversion as both halves are above the t-axis; it has a dc component of 4. But if this is removed so that the curve becomes symmetrical about the t-axis, the curve then has half-wave inversion. Its series has no even terms. Note that shifting a curve to left or right has no effect on its half-wave inversion. The wave of Figure 15.2d has a different type of symmetry in that the curve of the second half of the cycle, if *rotated* about the point $(\pi,0)$, lies exactly on the curve of the first half. However, this is not half-wave inversion; note that this curve has an *even* term in its series. We deal with curves of this type in the next paragraph.

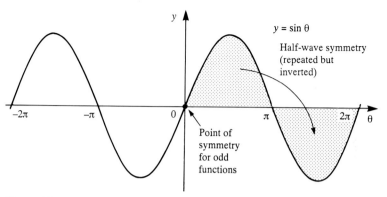

Figure 15.12

Two important types of symmetry are shown by odd and even functions. An **odd function** is symmetrical about the origin. In other words, if the curve is rotated 180° about the origin, it lies on itself. A sine curve (Figure 15.12) is a typical example. Examples in this chapter are Figures 15.4a and 15.11a, as well as the example of Figure 15.2d mentioned above. An odd function has sine terms only. Also, since the curve must inevitably be symmetrical about the t-axis, curves of this kind have no initial term.

An **even function** is one which is symmetrical about the y-axis. In other words, the portion to the left of the y axis is the mirror-image of the portion to the right. The cosine curve ($y = \cos t$, Figure 15.13) is a typical example of this. Another example is shown in Figure 15.11b. As might be expected, the series for such curves includes cosine terms but no sine terms. In other words, a_n exists but $b_n = 0$. The initial term, a_0 may be absent or present, depending on whether the function is symmetrical about the t-axis (Figure 15.13) or not (Figure 15.4b).

Figure 15.14 shows a curve which is not symmetrical (so is not properly included under this heading) but which has the feature known as **half-wave repetition**. The function repeats with a period of π instead of 2π. Functions of this kind may have both cosine and sine terms, but only those for even values of n.

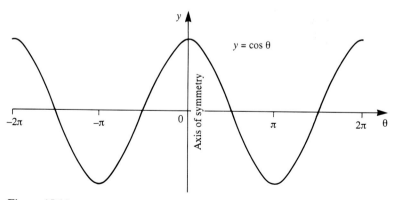

Figure 15.13

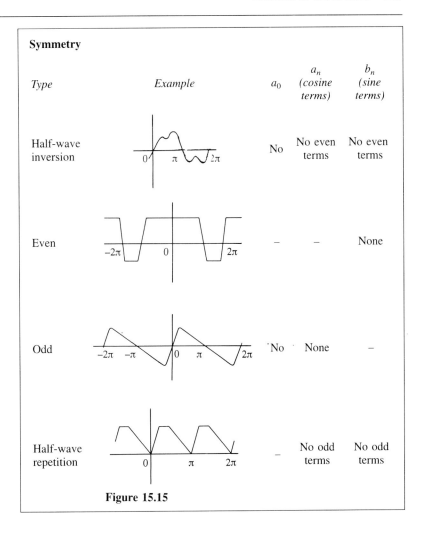

Figure 15.15

Functions may show more than one type of symmetry. The effects of symmetry in *eliminating* terms are cumulative. For example Figure 15.11a has both odd symmetry and half-wave inversion. Because of its odd symmetry it has *no* cosine terms. Because of its half-wave inversion is has *no* dc term and *no* even terms. The series consists of only odd sine terms.

Rectified ac

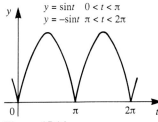

Figure 15.14

Alternating current, as supplied from a generator, consists of a more-or-less pure sine wave of a single frequency, the fundamental. After it has been rectified by a diode bridge, the waveform is as shown in Figure 15.14 The piecewise function is:

$y = \sin t \quad 0 < t < \pi$

$y = -\sin t \quad \pi < t < 2\pi$

Examination of the graph shows that this has half-wave repetition and even symmetry. The series therefore has no odd terms and no sine terms, consisting

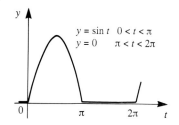

Figure 15.16

only of a dc term and even *cosine* terms. Rectification has produced a drastic change in the original pure *sine* wave.

It is interesting to analyse the two halves of the cycle separately, especially since this gives us the series for an ac wave which has been half-wave rectified (Figure 15.16). This wave has no symmetry, so we shall need to calculate a_0, a_n and b_n.

$$a_0 = \frac{1}{\pi} \int_0^\pi \sin t \, dt = \frac{1}{\pi} [-\cos t]_0^\pi = \frac{2}{\pi}$$

the initial term is $a_0/2 = 1/\pi$

$$a_n = \frac{1}{\pi} \int_0^\pi \sin t \cos nt \, dt$$

In this example, the usual technique of integrating by parts does not work, for it simply produces another integrand containing sine and cosine terms. Instead, we rely on one of the trig identities (page 68):

$$2 \sin A \cos B \equiv \sin (A + B) + \sin (A - B)$$

In this case, put $A = t$ and $B = nt$, and split the integral into two:

$$a_n = \frac{1}{2\pi} \left\{ \int_0^\pi \sin(1+n)t \, dt + \int_0^\pi \sin(1-n)t \, dt \right\}$$

$$= \frac{-1}{2\pi} \left[\frac{\cos(1+n)t}{1+n} + \frac{\cos(1-n)t}{1-n} \right]_0^\pi$$

The negative sign from integrating the sines is taken outside the bracket. Adding the fractions:

$$= \frac{-1}{2\pi} \left[\frac{(1-n)\cos(1+n)t + (1+n)\cos(1-n)t}{(1+n)(1-n)} \right]_0^\pi$$

Multiplying out and then rearranging terms:

$$= \frac{-1}{2\pi} \left[\frac{\{\cos(1+n)t + \cos(1-n)t\} - n\{\cos(1+n)t - \cos(1-n)t\}}{1 - n^2} \right]_0^\pi$$

Now we use two more trig identities to replace the first pair of terms and the last pair of terms:

$$2 \cos A \cos B \equiv \cos(A + B) + \cos(A - B)$$

$$2 \sin A \sin B \equiv \cos(A + B) - \cos(A - B)$$

$$a_n = \frac{-1}{\pi} \left[\frac{\cos t \cos nt + \sin t \sin nt}{1 - n^2} \right]_0^\pi$$

The sine terms all equal zero, so this simplifies to:

$$a_n = \frac{-[-\cos n\pi - 1]}{\pi(1 - n^2)} = \frac{\cos n\pi + 1}{\pi(1 - n^2)}$$

If n is even, then $a_n = \dfrac{2}{\pi(1 - n^2)}$

If n is odd, then $a_n = 0$.

However, there is one odd number not yet accounted for. When $n = 1$, the denominator becomes zero, so the expression is indeterminate. We must evaluate this separately:

$$a_1 = \frac{1}{\pi} \int_0^\pi \sin t \cos t \, dt = \frac{1}{2\pi} \int_0^\pi \sin 2t \, dt = 0$$

The calculation of b_n follows steps similar to those above, with the result that $b_n = 0$. Again the general expression is indeterminate for $n = 1$. Calculation shows that $b_n = 1/2$.

From these results we obtain the series:

$$y = \frac{1}{\pi} \left\{ 1 - \frac{2}{3} \cos 2t - \frac{2}{15} \cos 4t - \frac{2}{35} \cos 6t - \ldots + \frac{\pi}{2} \sin t \right\}$$

The cosine terms are negative because $(1 - n^2)$ is negative when $n > 1$. The series for the second half of the cycle (Figure 15.17) gives an almost identical result:

$$y = \frac{1}{\pi} \left\{ 1 - \frac{2}{3} \cos 2t - \frac{2}{15} \cos 4t - \frac{2}{35} \cos 6t - \ldots - \frac{\pi}{2} \sin t \right\}$$

The only difference is the sign of the last term. When these two series are added to obtain the series for the full-wave rectified wave-form (Figure 15.14), the two sine terms cancel each other, and the other terms are doubled:

$$y = \frac{2}{\pi} \left\{ 1 - \frac{2}{3} \cos 2t - \frac{2}{15} \cos 4t - \frac{2}{35} \cos 6t - \ldots \right\}$$

This consists of $a_0/2$ and the even cosine terms, as predicted at the start of this discussion. This is a series which converges rapidly, because of n^2.

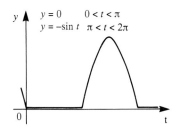

Figure 15.17

Test yourself 15.2

1 Complete the calculations of the Fourier series for the function of Figure 15.15 by finding the values of b_n and b_1 given above.

2 For each of the wave-forms of Figure 15.18, state which kind or kinds of symmetry it shows, and which groups of terms are *not* present in its Fourier series.

3 Given the function:
$$y = \pi - t \quad 0 < t < \pi$$
$$y = -\pi + t \quad \pi < t < 2\pi$$
sketch the function, list its symmetries, state which groups of terms are absent from its Fourier series and calculate the first four terms of the series.

4 Given the function:

$$y = 3 \qquad 0 < t < \pi/2$$
$$y = 0 \qquad \pi/2 < t < 3\pi/2$$
$$y = 3 \qquad 3\pi//2 < t < 2\pi$$

sketch the function, list its symmetries, state which groups of terms are absent from its Fourier series and calculate the first five terms of the series.

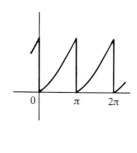

a

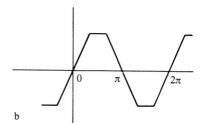

b

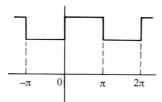

c

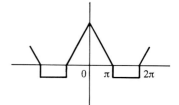

d

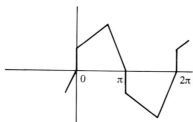

e

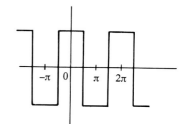

f

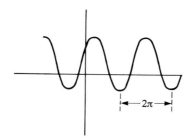

g

h

Figure 15.18

Explore this

Use a calculator or a micro with a graph-plotting program to investigate Fourier series. Plot graphs of the series for the functions listed in the text and 'Test yourself' 15.1 and 15.2. Test the effect of increasing the number of terms summed. Derive the series for other functions and plot these too.

Answers

Chapter 1

Try these first (page 3)

1. **a** All
   b 1.2, 2, –6, ⅖, $-\dfrac{4}{7}$, –0.7, 73
   c 2, 73
   d 2, –6, 73
2. 7, 2, 13
3. 2, 11, 13
4. 6
5. 120
6. **a** $(2 + x)(2 - x)$ **b** $3ab^3(b + 2a)$ **c** $(2x - 5)(x - 2)$
7. **a** 2⁷⁄₁₅ **b** 5¾ **c** 1⅖
8. **a** $\dfrac{3}{2a + 6}$ **b** $\dfrac{3x}{14z}$ **c** $\dfrac{6x}{(x - 2)(2x + 1)}$
9. $\dfrac{2}{x + 2} + \dfrac{5}{x - 3}$
10. **a** x^7 **b** a^{12} **c** b^4
11. 5
12. 3
13. 4.2, 5.38, 21.74, 7.50, 35.44
14. $|\mathbf{A}| = 5$, $\theta_A = 36.87°$
    $|\mathbf{B}| = 7.28$, $\theta_B = 105.95°$
    $\mathbf{C} = \begin{vmatrix} 2 \\ 10 \end{vmatrix}$, $|\mathbf{C}| = 10.20$, $\theta_C = 78.69°$

Test yourself 1.1

1. 3, –5, 0, –100
2. All except 2π
3. All
4. 56
5. 1, 2, 3, 6, 7, 14, 21, 42
6. 1, 3, 5, 9, 15, 45
7. 1, 5, 7, 35, 55, 77, 385
8. 1, 2, 3, 5, 6, 10, 13, 15, 25, 26, 30, 39, 50, 65, 75, 78, 130, 150, 195, 325, 390, 650, 975, 1950
9. 10
10. 21
11. 39
12. 6
13. 2310
14. 840
15. 720
16. 13 860

Test yourself 1.2

1. $x^2(5x + 1)$
2. $2ab(b + a)$
3. $(a + 3)(a - 3)$
4. $(4 + a)(x + y)$
5. $(x + 4)(x + 2)$
6. $(x - 5)(x + 2)$
7. $(x - 5)(x - 3)$
8. $(p + a)(5 + p)$
9. $m(m - 5)$
10. $(x - 3)(x + 5)$
11. $(3x - 5)(5x - 4)$
12. $6(x + y)(x + 3)$
13. $(x + 2)(x - 3)$
14. $5(x + 4)(x + 1)$
15. $(3x + 4)(x - 2)$
16. $(3x + 4)(3x - 4)$
17. $(2a + b)(x - 3y)$
18. $3(12 - x)(1 + x)$
19. $(x - 6)(x - 7)$
20. $(x + 7)(x - 5)$
21. $(2x + 3)(x + 5)$
22. $(3m + 1)(3m - 1)$
23. $(2x + 3)(3x + 4)$
24. $(4x - 5)(2x + 3)$

Chapter 2

Try these first (page 12)

1. a $2\frac{7}{15}$ b $5\frac{3}{4}$ c $1\frac{2}{5}$
2. a $\dfrac{3}{2a + 6}$ b $\dfrac{3x}{14z}$ c $\dfrac{6x + 8}{(x - 2)(2x + 1)}$
3. $\dfrac{2}{x + 2} + \dfrac{5}{x - 3}$

Test yourself 2.1

1. $1\frac{1}{12}$
2. $\dfrac{11}{20}$
3. $3\frac{3}{10}$
4. $1\frac{1}{8}$
5. $\dfrac{5}{8}$
6. $\dfrac{21}{32}$
7. $1\frac{1}{5}$
8. $1\frac{1}{3}$
9. $\dfrac{5}{8}$
10. $\dfrac{10}{27}$

Test yourself 2.2

1. $\dfrac{2}{3x + 1}$
2. $\dfrac{3}{x + 2}$
3. x
4. $\dfrac{5}{3}$
5. $\dfrac{x + 3}{x - 4}$
6. $\dfrac{x - 5}{x - 3}$

7 $\dfrac{8xy^2}{7z}$ 8 $\dfrac{6x^2 - 3x}{x^2 - 4}$

9 $\dfrac{3x}{10}$ 10 $\dfrac{5x + 3}{4y}$

11 $\dfrac{2b + 3}{ab}$ 12 $\dfrac{3x - 2xyz}{y^2 z}$

13 $\dfrac{2(4x + 7)}{(x + 1)(x + 3)}$ 14 $\dfrac{(x + 5)}{(x + 1)(x + 2)}$

15 $\dfrac{5x - 1}{x^2 - 1}$ 16 $\dfrac{3 - x - x^2}{(x + 1)^2}$

Test yourself 2.3

1 $\dfrac{2}{x - 3} + \dfrac{3}{x - 7}$ 2 $\dfrac{5}{x - 1} - \dfrac{1}{x + 4}$

3 $\dfrac{17}{7(x - 4)} + \dfrac{4}{7(x + 3)}$ 4 $\dfrac{1}{x + 2} - \dfrac{4}{(x + 2)^2} + \dfrac{5}{(x + 2)^3}$

Chapter 3

Try these first (page 23)

1 **a** x^7 **b** a^{12} **c** b^4
2 5 3 3
4 $|\mathbf{A}| = 5$, $\theta_A = 36.87°$
 $|\mathbf{B}| = 7.28$, $\theta_B = 105.95°$
 $\mathbf{C} = \begin{pmatrix} 2 \\ 10 \end{pmatrix}$, $|\mathbf{C}| = 10.20$, $\theta_C = 78.69°$

Test yourself 3.1

1 a^6 2 5
3 a^6 4 $9a^6$
5 $8p^8$ 6 $x^5/3$
7 a^2 8 n^0 (=1)
9 a^2 10 $3b^2$

Test yourself 3.2

1 **a** 2 **b** 3 **c** 6
2 **a** $\log_2 8 = 3$ **b** $\log_5 625 = 4$
 c $\log_3 1 = 0$ **d** $\log_{36} 6 = 0.5$
 e $\log_5 0.008 = -3$ **f** $\log_{25} 0.2 = -0.5$

ANSWERS 295

3 **a** $2^6 = 64$ **b** $3^0 = 1$
   c $5^1 = 5$ **d** $4^3 = 64$

4 **a** log 8   **b** log 5
   c log 9   **d** log 4
   e log 147   **f** log 80

5 **a** 3   **b** 3
   c 1.161   **d** –1.585

Test yourself 3.3

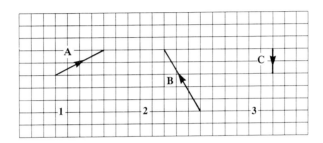

4 $\mathbf{D} = \begin{pmatrix} 6 \\ 7 \end{pmatrix}$ **5** $\mathbf{E} = \begin{pmatrix} -6 \\ -1 \end{pmatrix}$ **6** $\mathbf{F} = \begin{pmatrix} -5 \\ 4 \end{pmatrix}$

7 $|\mathbf{A}| = 4.24, \theta = 315.00°$

8 $|\mathbf{B}| = 4.12, \theta = 14.04°$

9 $|\mathbf{C}| = 7.81, \theta = 129.81°$

10 $|\mathbf{D}| = 8.06, \theta = 209.74°$

11 $|\mathbf{E}| = 5.83, \theta = 329.04°$

12 $|\mathbf{F}| = 7.28, \theta = 74.05°$

13 $\mathbf{A} + \mathbf{B} = \begin{pmatrix} 2 \\ 6 \end{pmatrix}, |\mathbf{A} + \mathbf{B}| = 6.32, \theta = 71.57°$

14 $\mathbf{A} - \mathbf{B} = \begin{pmatrix} 4 \\ 2 \end{pmatrix}, |\mathbf{A} - \mathbf{B}| = 4.47, \theta = 26.57°$

15 $\mathbf{A} + \mathbf{C} = \begin{pmatrix} 5 \\ 6 \end{pmatrix}, |\mathbf{A} + \mathbf{C}| = 7.81, \theta = 50.19°$

16 $\mathbf{B} + \mathbf{D} = \begin{pmatrix} -4 \\ 0 \end{pmatrix}, |\mathbf{B} + \mathbf{D}| = 4, \theta = 180°$

17 $\mathbf{A} + \mathbf{C} + \mathbf{D} = \begin{pmatrix} 2 \\ 4 \end{pmatrix}, |\mathbf{A} + \mathbf{C} + \mathbf{D}| = 4.47, \theta = 63.43°$

18 $\mathbf{B} + \mathbf{D} - \mathbf{C} = \begin{pmatrix} -6 \\ -2 \end{pmatrix}, |\mathbf{B} + \mathbf{D} - \mathbf{C}| = 6.32, \theta = 198.43°$

19 y-axis = 2.39, x-axis = 6.58

20 y- and x-axis = 2.83

21 y-axis = 2.30, x-axis = –1.93

Chapter 4

Try these first (page 33)

1. $R_B = \dfrac{V_{BE}\, R_A}{V - V_{BE}}$
2. **a** $x = 6$ or $x = -3$ **b** $x = 0.71$ or $x = -4.21$
3. **a** $x = 7, y = -4$ **b** $x = -7, y = -3$ or $x = 1.5, y = 14$
4. $x = -1, y = 10, z = 3$
5. **a** 75 000 **b** 25.75
6. **a** 5.370 **b** 1.32
7. **a** 2.473×10^3 **b** 5.2×10^{-2}
8. **a** 3.26×10^3 **b** 5.3742×10^{-6}

Test yourself 4.1

1. $\dfrac{E - V}{r}$
2. $\dfrac{3v}{\pi r^2}$
3. $\dfrac{Pq}{m}$
4. $\dfrac{c + 3u}{v}$
5. $\dfrac{t^2 + k}{2}$
6. $\pm \sqrt{\dfrac{y - b}{a}}$
7. $\dfrac{x}{4a^2}$
8. $\pm \sqrt{(2x + 14)}$
9. $\dfrac{R_1 R_2}{R_1 + R_2}$
10. $\dfrac{by}{b + 2y}$
11. $\dfrac{5 - 4x}{x - 2}$
12. $\dfrac{S - \pi r^2}{2\pi r}$

Test yourself 4.2

1. $-2, 4$
2. $1.5, -1.25$
3. $2, -6$
4. $3, 1$
5. $-2, 5$
6. $3, -3$
7. $3.83, -1.83$
8. $0.77, -0.43$
9. $1.39, 0.36$
10. $0.87, -2.87$

Test yourself 4.3

1. $x = 2, y = 3$
2. $x = 4, y = 1$
3. $x = -2, y = 5$
4. $x = 7, y = -4$
5. $x = 3, y = 2$
6. $x = -7, y = -3$
7. $x = 1, y = -1$
8. $x = \frac{1}{2}, y = 7$
9. $x = -3, y = 8$
10. $x = 3, y = -2$
11. $x = 0, y = 1$
    $x = 2, y = 3$
12. $x = 3, y = 5$
    $x = -7, y = -1$
13. $x = 3, y = -2, z = 1$
14. $x = 0, y = 1, z = -1$

Test yourself 4.4

1. **a** 46 **b** 3204.7
   c 35 200 **d** 0.0076
   d 6.0 **e** 800
2. **a** 4.329 **b** 5.66
   c 7.7 **d** 0.0074
3. **a** 1.429×10^3 **b** 2.3470×10^4
   c 7.5×10^{-2} **d** 1×10^{-7}
   e 2.46×10^3 **f** 4.736×10^{-5}
4. **a** 1.429×10^3 **b** 23.470×10^3
   c 75×10^{-3} **d** 100×10^{-9} (1 sf)
   e 2.46×10^3 **f** 47.36×10^{-6}

Chapter 5

Try these first (page 48)

1. **a** Straight line, sloping up to the right
   b 1.5
   c $-3\frac{1}{2}$
   d $-2\frac{1}{3}$
   e No
   f 5.41 (2 dp)
2. **a** Parabola, $y = (x - 3)^2$
   b Exponential, $y = 5e^x$
   c Linear, $y = 7 - 2x$
   d Exponential, $y = 4x^3$
   e Exponential, $y = 5e^{2x}$
3. **a** 39
   b 4
4. **a** B
   b E
   c A, $x = 3$
   d C

Test yourself 5.1

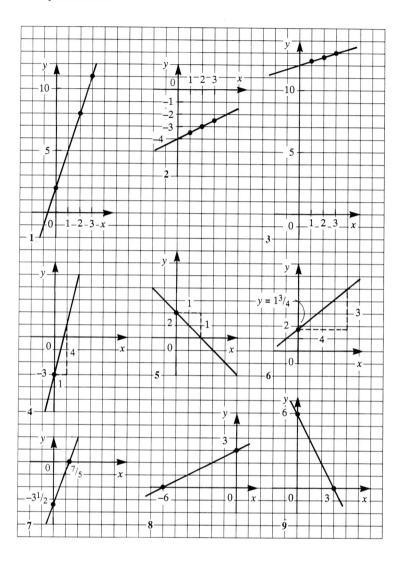

Test yourself 5.2

1. $m = 1$, $c = 2$, $y = x + 2$
2. $m = \dfrac{-1}{2}$, $c = 5$, $y = 5 - \dfrac{x}{2}$, or $2y = 10 - x$
3. $m = 3$, $c = 3$, $y = 3x + 3$
4. $m = \dfrac{2}{3}$, $c = -2$, $y = \dfrac{2x}{3} - 2$, or $3y = 2x - 6$
5. $m = 1$, $c = -3$, $y = x - 3$
6. $m = -2$, $c = -7$, $y = -2x - 7$

7 $m = 0$, $c = 3$, $y = 3$
8 $m = 2$, $c = 1$, $y = 2x + 1$
9 $m = -3$, $c = 8$, $y = 8 - 3x$
10 Yes   11 No   12 No
13 Yes   14 No   15 No
16 1.414   17 4.243   18 11.314
19 10.296

Test yourself 5.3

1 Parabola, $y - 5 = x^2$
2 Exponential, $y = e^{x-5}$
3 Straight line, $y = \dfrac{-x}{2} - 5$
4 Parabola, $y - 7 = (x + 5)^2$
5 Exponential, $y = -e^x$
6 Parabola, $y = -(x + 7)^2$

Test yourself 5.4

1 Straight line on semi-log paper; $a = 2$, $b = 0.5$
2 Straight line on log-log paper; $k = 3$, $n = 1.2$
3 Straight line on log-log paper; the equation is $y = 0.5x^{3.5}$

Test yourself 5.5

In question 1, did you remember to select the correct mode, degrees or radians?

1 a 0.0000   b 0.2588   c −1.0000
  d −1.0000   e −1.3764   f −0.8011
  g 0.8481 rad and 2.2935 rad (48.5903° and 131.4097°)
  h 2.2916 rad and 3.9916 rad (131.2999° and 228.7001°)
  i 1.1071 rad and 4.2487 rad (63.4349° and 243.4349°)
2 a $x = 3$, $y = 6$
  b $x = 1$, $y = 3$ and $x = 4$, $y = 18$
3 a −3, 5, 197
  b 0.054 95, 3.000, 4.883×10^5
  c 1.438, 2.847, −0.8382
4 a, c, d, e, f, g

Chapter 7

Test yourself 7.1

1 a A(7, 40°)   **b** B(5, 100°)
   c C(4, 10°)   **d** D(6, −50°)
   e E(3, −135°)

2

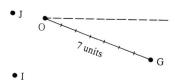

3 a P(3, 60°)   **b** P(3, 10°)   **c** 100° s^{-1}
4 50π rad s^{-1}
5 A = 7$\underline{|40°}$, B = 5$\underline{|100°}$, C = 4$\underline{|10°}$, D = 7$\underline{|−50°}$,
   or D = 6$\overline{|50°}$, E = 3$\underline{|−135°}$, or E = 3$\overline{|135°}$

Test yourself 7.2

1 $y_P = 3 \sin \omega t$
   $y_Q = 4 \sin(\omega t + 50°)$
   $y_R = 2 \sin(\omega t − 45°)$
2 $y_L = 2 \sin \omega t$
   $y_M = 3 \sin(\omega t + 120°)$
   $y_N = 2.5 \sin(\omega t + 190°)$
   $y_N = 2.5 \sin(\omega t − 170°)$

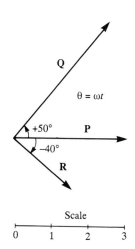

Test yourself 7.3

1 a A(8.0, 60.3°)   **b** B(6.7, 116.6°)
   c C(9.2, −192.5°)   **d** D(7.1, −8.1°)
2 a A(27.7, 11.2)   **b** B(−7.9, 9.1)
   c C(−15, 0)   **d** D(39.4, −23.7)

Chapter 8

Test yourself 8.1

1. 14
2. 2
3. No limit
4. 6
5. 1/4
6. 0
7. 3
8. 9

Test yourself 8.2

1. −1
2. 1 (*any* number to power 0 equals 1)
3. 3
4. 4/5
5. 3/2
6. 2
7. 5

Test yourself 8.3

1. 0
2. 1
3. 0 (The value oscillates above and below 0, gradually getting closer to 0.)

Chapter 9

Test yourself 9.1

1. Substituting $x = 5$ in $dy/dx = 2x + 5$, $m = 10 + 5 = 15$
2. Substituting $x = 0$ in $dy/dx = 2x − 11$, $m = 0 − 11 = −11$
3. Substituting $q = −3$ in $dp/dq = 2q + 4$, $m = −6 + 4 = −2$
4. Substituting $x = 1$ in $dy/dx = 6x + 2$, $m = 6 + 2 = 8$
5. Substituting $x = 7$ in $dy/dx = 3x^2 + 2x + 1$, $m = 147 + 14 + 1 = 162$
6. Substituting $h = 2$ in $dg/dh = 12h^2 − 10h + 13 = 48 − 20 + 13 = 41$

Test yourself 9.2

1. $4x^3$
2. 0
3. $735x^{20}$
4. −8
5. $3\pi x^2$
6. $12x^2 − 2$
7. $5.2x^3$
8. px^{p-1}
9. k
10. $\dfrac{-2}{x^2}$
11. $\dfrac{-8}{x^3}$
12. $\dfrac{1}{\sqrt{x}}$
13. $4 + \dfrac{1}{x^2}$
14. $8x + 1 + \dfrac{8}{x^5}$
15. $\dfrac{1}{3 \cdot \sqrt[3]{x^2}}$ or $\dfrac{x^{-2/3}}{3}$

16. **a** 1.25 mV **b** 0 V **c** −5 mV
17. **a** −20 mV **b** −40 mV
    c $i = 0$ when $t = \sqrt{0.1} = 0.316\,\text{s}$, and then $v = −63.2\,\text{mV}$

Test yourself 9.3

1 $20x^9$ **2** $\dfrac{-12}{x^{10}}$

3 $\cos\theta = 0.8090$ (4 dp)   **4** $-2\sin\theta = 0$
5 $e^x = 20.09$ (4 dp)   **6** $4/x = 1$
7 $3^x (\ln 3)$ **8** $1/x$
9 $4\cos 4\theta$ **10** $3e^{3x}$
11 $-8\sin 2\theta$ **12** $6\cos 2\theta + \sin(\theta/2)$
13 $dv/dt = 20\cos 2t$; $i = 921\,\mu\text{A}$
14 $di/dt = 0.1\pi f \cos 2\pi ft$:   **a** -31.4 A s^{-1} **b** 63.5 mV

15 $\dfrac{di}{dt} = \dfrac{E}{L} e^{-Rt/L}$:   **a** 20 A s^{-1} **b** 0.996 A s^{-1}

16 $\dfrac{dv}{dt} = \dfrac{-V}{RC} e^{-t/RC}$:   **a** -25.5 V s^{-1} **b** -16.7 V s^{-1}

Test yourself 9.4

1 (2, 1) is a minimum
2 (3, 5) is a maximum
3 (−3, 167) is a maximum, (5, −345) is a minimum

Test yourself 9.5

1 $18x - 30$ **2** $15(3x+2)^4$
3 $15(x+3)^2$ **4** $2(2x^2 - 3x + 7)(4x-3)$

5 $\dfrac{-3x^2}{(x^3+2)^2}$ **6** $\dfrac{3}{2\sqrt{3x+2}}$

7 $5\cos 5\theta$ **8** $\dfrac{1}{x}$

Test yourself 9.6

1 $12x - 7$ **2** $8x^3 - 15x^2 + 2$
3 $16x^7 - 28x^6 - 18x^2 + 24x$ **4** $243x^8 + 189x^6 + 45x^4 + 3x^2$

5 $\dfrac{-3}{(x-2)^2}$ **6** $\dfrac{26}{(2x+3)^2}$

7 $\dfrac{-(2x+9)}{x^4}$ **8** $\dfrac{2x-1}{2x\sqrt{x}}$

9 $\dfrac{(2x+3)(2x-3)}{5x^2}$

Test yourself 9.7

1. $\partial z/\partial x = 10x + 2y$, $\partial z/\partial y = 2x - 2y$
2. $\partial z/\partial x = 18x^2 + 30xy^2 - 2y$
   $\partial z/\partial y = 30x^2y - 2x - 15y^2$
3. $\partial z/\partial x = 2 \cos(2x - 3y)$
   $\partial z/\partial y = -3 \cos(2x - 3y)$
4. $\partial P/\partial R = 2V/R$, $\partial P/\partial R = -V^2/R^2$, $\Delta P = 10.8$ (3 sf)
5. $\partial i/\partial v = -1/r$, $\partial i/\partial r = -\dfrac{V - v}{r^2}$, $di/dt = -7 \times 10^{-6}$ when the current is 0.046 making the tempco -152 ppm per °C

Chapter 10

Test yourself 10.1

1. $\dfrac{x^3}{3} - x^2 + c$
2. $\dfrac{x^4}{4} - \dfrac{3x^2}{2} + 2x + c$
3. $7x + c$
4. $\dfrac{x^3}{3} + \dfrac{5x^2}{2} + 6x + c$
5. $\dfrac{3x^2}{2} + 2x^3 + c$
6. $2\sqrt{x} + c$
7. $\dfrac{\sin 3x}{3} + c$
8. $\cos\theta + 3\theta + c$
9. $\dfrac{-2 \sin 3x}{3} + c$
10. $\dfrac{e^{3x}}{3} + c$
11. $\dfrac{e^{4x}}{4} - e^{2x} + c$
12. $\dfrac{7}{4} \ln(x) + c$ or $\dfrac{7}{4} \ln(kx)$
13. $x^2 + 4x + 1$
14. $x^3 + 2x^2 - 15x + 3$
15. $\dfrac{-1}{x} + \dfrac{1}{3x^3} + 2$
16. 27
17. 37.33
18. 64
19. 2
20. 4.667
21. $\dfrac{e^2 - 1}{2} = 3.195$

Test yourself 10.2

1. $2x - \dfrac{7}{2} \ln|2x + 1| + c$
2. $\dfrac{3x^2}{2} + 9x + 26 \ln|x - 3| + c$
3. $x^2 - 8 \ln|x + 3| + c$
4. $\dfrac{(2x + 1)^6}{12} + c$

5 $x^3 + \dfrac{x^2}{2} + c$ **6** $x + 1 - \ln|x + 1| + c$

7 $-x \cos x + \sin x + c$ **8** $\dfrac{x^2}{2} \ln x - \dfrac{x^2}{4} + c$

9 $e^x (x - 1)$

Test yourself 10.3

1 9   **2** 5
3 a 6   **b** 12   **c** 12
4 1.667, 1.461   **5** 12.724, 16.379
6 −0.673, 0.857   **7** 0.212, 0.707
8 2.187, 2.489   **9** 212.1
10 339.46

Chapter 11

Test yourself 11.1

1 $y = 3x + c$, $c = 1$, $y = 28$
2 $y = 5x + c$, $c = 2$, $y = 7$
3 $y = Ae^{3x}$, $A = 1$, $y = 8103$
4 $y = Ae^{-4x}$, $A = 163.8$, $y = 22.2$
5 a 32.2 s   **b** 46.1 s
6 a 2.57 A   **b** 1.89 m s

Test yourself 11.2

1 a $i = \dfrac{m}{R^2}[Rt + L(e^{-Rt/L} - 1)]$ **b** 5.62 μA

2 a $R \cdot \dfrac{dq}{dt} + \dfrac{q}{C} = v$ **b** $\dfrac{dq}{dt} + \dfrac{1}{RC} \cdot q = \dfrac{v}{R}$

c $e^{t/RC}$ **d** $q = e^{-t/RC} \int \dfrac{v}{R} e^{t/RC} \, dt + Ae^{-t/RC}$

e $q = Cv + Ae^{-t/RC}$ **f** $A = -Cv$; $q = Cv(1 - e^{-t/RC})$
g $v_c = q/C = v(1 - e^{t/RC})$ **h** $v = 1.70$ V

3 $f(x) = 4$; $F(x) = 4x$; int. fact. $= e^{4x}$; $y = Ae^{-4x}$
4 $f(x) = -b$; $F(x) = -bx$; int. fact. $= e^{-bx}$; $y = Ae^{bx}$
5 $f(x) = 1/x$; $F(x) = \ln x$; int. fact. $= e^{\ln x} = x$; $y = Ae^{-\ln x} = A/x$

6 $f(x) = 2$; $g(x) = x$; $F(x) = 2x$; int. fact. $= e^{2x}$; $G(x) = \int e \cdot 2^x \, dx$
$= e^{2x}(x/2 - 1/4)$; $y = x/2 - 1/4 + Ae^{-2x}$

7 $f(x) = 4$; $g(x) = e^x$; $F(x) = 4x$; int. fact. $= e^{4x}$; $G(x) = e^{5x}/5$;
$y = e^x/5 + Ae^{-4x}$

8 $f(x) = 1$; $g(x) = x^2 e^{-x}$; $F(x) = x$; int. fact. $= e^x$; $G(x) = x^3/3$;
$$y = \frac{x^3 e^{-x}}{3} + Ae^{-x}$$

Test yourself 11.3

1 $y = Ae^{5x} + Be^x$ **2** $y = Ae^{3x} + Be^{-4x}$
3 $y = Ae^x \cos 2x + Be^x \sin 2x$ **4** $y = Ae^{-2x} + Bxe^{-2x}$
5 $y = A + Be^{-5x}$ **6** $y = Ae^{5x} + Be^{-2x}$; $y = e^{5x} - e^{-2x}$
7 $y = Ae^{2x} + Be^x$; $y = -3e^{2x} + 4e^x$
8 $y = Ae^{4x} \cos 3x + Be^{4x} \sin 3x$; $y = 2e^{4x} \cos 3x - e^{4x} \sin x$
9 $i = 1.033 \times 10^{-5} e^{-2764t} - 5.33 \times 10^{-6} e^{-7236t}$; when $t = 300$ μs,
$i = 3.90$ μA

10 For $D = 0$ we make $f^2 = 4g \Rightarrow R = 2\sqrt{\dfrac{L}{C}}$. $R = 975.9 \, \Omega$;
$y = e^{-4880t} + 4880te^{-4880t}$

11 The current is reduced to zero as fast as possible but without oscillating.

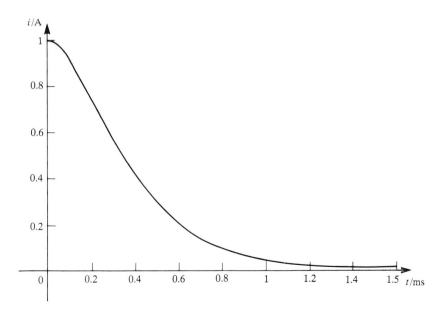

Chapter 12

Test yourself 12.1

1 By Euler: $y = 48$, $y = 76.9$, $y = 106.6$. Solution is $y = 3e^x$ (this is a growth equation, page 160). Exact solution: $y = 163.8$.
2 By Euler: $y = 6.78$. Solution is $y = x^3 - 5$. Exact solution: $y = 6.852$.
3 By Euler: $y = 8.44$. Exact solution: $y = 9.59$.

Test yourself 12.2

1 Ae^{-5x} **2** $Ae^{-3x} + Be^{2x}$
3 $Ae^{4x} + Be^{2x}$ **4** $Ae^{-2x} + Bxe^{-2x}$
5 $Ae^{-4x} + Be^{4x}$ **6** $A\cos 4x + B\sin 4x$
7 $e^x(A\cos 2x + B\sin 2x)$ **8** $Ae^{2x} + Be^{-5x} + Ce^{-x}$
9 $Ae^{-x} + Be^x + C\cos 2x + D\sin 2x$
10 $e^{-2x}(A\cos 3x + B\sin 3x)$

Test yourself 12.3

1 $\dfrac{-5}{s}$ **2** $\dfrac{3}{s^2}$ **3** $\dfrac{1}{s+2}$

4 $\dfrac{s}{s^2+9}$ **5** $4t$ **6** $3e^{-7t}$

7 $e^{-2t}\cos 5t$ **8** $e^{-3t}\cos 4t$ **9** $1 - e^{-3t}$

10 $\sin(3t+2)$ **11** $\sin 2t + \cos 2t$ **12** $2e^{2t} + 3e^{-t}$

13 $5t - 2e^{4t}$ **14** $V(s) = \dfrac{2}{s+5}$; $v(t) = 2e^{-5t}$

15 $I(s) = \dfrac{1}{s-2} - \dfrac{1}{s+3}$; $i(t) = e^{2t} - e^{-3t}$

16 $I(s) = \dfrac{2}{s+2}$; $i(t) = 2e^{-2t}$

17 $I(s) = \dfrac{2}{s+0.25}$; $i(t) = 2e^{-0.25t}$

18 a $\dfrac{v}{5} + 0.1\dfrac{dv}{dt} = 2$

    b $V(s) = \dfrac{20}{s(s+2)}$

    c $v = 10(1 - e^{-2t})$

19 a $2i(t) + 0.1 \dfrac{di(t)}{dt} = 2t$

b $I(s) = \dfrac{2}{(2 + 0.1\,s)s^2}$

c $i(t) = 0.05\,(e^{-20t} - 1) + t$

20 a $2i(t) + \dfrac{1}{0.1}\displaystyle\int_0^t i(T)\,dT = 4$

b $I(s) = \dfrac{2}{s + 5}$

c $i(t) = 2e^{-5t}$

21 a $i(0) = 10$ A; $i(\infty) = 0$ A
b $i(0) = 0$ A; current increases indefinitely with time.
c $i(0) = 2$ A; $i(\infty) = 0$ A

22 $i(0) = 0$ A; $i(\infty) = 1.67$ A

23 $i(0) = 0$ A; $i(\infty) = 1.5$ A

Chapter 13

Test yourself 13.1

1 1	**2** 7	**3** −11
4 19	**5** −46	**6** 23

Test yourself 13.2

1 $x = 3, y = 2$ **2** $x = 4, y = -1$
3 $x = -2, y = -5$ **4** $x = 7, y = 3$
5 $x = 2, y = 1/5$
6 $i_1 = 2.\dot{0}\dot{9}$ A (or $2\tfrac{1}{11}$ A), $i_2 = -1.\dot{2}\dot{7}$ A (or $-1\tfrac{3}{11}$ A). The current through the 2Ω resistor is $i_1 - i_2 = 3.\dot{3}\dot{6}$ A (or $3\tfrac{4}{11}$ A) in the same direction as i_1.

Test yourself 13.3

2 6
3 94. Choosing the middle column or bottom row, so that one of the elements is 0, means that only two co-factors have to be evaluated, instead of three.
4 a $x = 2, y = 5, z = 1$
  b $x = 4, y = -2, z = 3$. Choose the first column or bottom row for easier evaluation.

Test yourself 13.4

1 −10   **2** 145   **3** 36
4 0   **5** −62
6 $x = -4, y = 3, z = 5$
7 $x = 2, y = -5, z = -1$
8 $x = 3, y = 0, z = -5$
9 $w = 2, x = -2, y = -1, z = 3$
10 $i_1 = 0.2$ A, $y = 0.1$ A, $z = 0.1$ A
11 $i_1 = 5$ A, $i_2 = 19/3$ A, $i_3 = 17/3$ A

Test yourself 13.5

$\mathbf{Ab} = \begin{bmatrix} 10 \\ 5 \end{bmatrix}$ $\mathbf{bc} = \begin{bmatrix} -6 & 3 \\ -8 & 4 \end{bmatrix}$ $\mathbf{cb} = \begin{bmatrix} -2 \end{bmatrix}$

$\mathbf{A}^2 = \begin{bmatrix} 7 & 1 \\ 3 & 2 \end{bmatrix}$ $\mathbf{cA} = \begin{bmatrix} -1 & -3 \end{bmatrix}$ $\mathbf{DE} = \begin{bmatrix} 5 & 5 \\ 0 & 3 \\ 11 & 3 \end{bmatrix}$

$\mathbf{EA} = \begin{bmatrix} 8 & -1 \\ -6 & 2 \\ 9 & 2 \end{bmatrix}$

It is not possible to calculate **Ac**, **bD**, **ED** and **AE**

Test yourself 13.6

1 $\dfrac{\begin{bmatrix} 7 & -5 \\ -3 & 4 \end{bmatrix}}{13}$ **2** $\dfrac{\begin{bmatrix} 1 & 5 \\ -3 & 2 \end{bmatrix}}{17}$

3 Determinant is zero, no inverse.

4 $\dfrac{\begin{bmatrix} 7 & -5 \\ 2 & -3 \end{bmatrix}}{-11}$

5 $\dfrac{\begin{bmatrix} -1 & 1 & 1 \\ 1 & -1 & 1 \\ 1 & 1 & -1 \end{bmatrix}}{2}$ **6** $\dfrac{\begin{bmatrix} 0 & 2 & -1 \\ -3 & 0 & 3 \\ 3 & -1 & -1 \end{bmatrix}}{3}$

7 Determinant is zero, no inverse.

8 $\dfrac{\begin{bmatrix} 10 & -1 & 2 \\ -2 & 8 & 10 \\ 4 & -3 & 6 \end{bmatrix}}{26}$

9 $\dfrac{\begin{bmatrix} 7 & 1 & -2 \\ 28 & 1 & -2 \\ -5 & -1 & 1 \end{bmatrix}}{-1}$

Test yourself 13.7

1 $x_1 = 1, x_2 = 3, x_3 = 2$
2 $x_1 = 2, x_2 = 3, x_3 = -3$
3 $x_1 = 3, x_2 = 0, x_3 = -1$

Chapter 14

Test yourself 14.1

1 $j, j2, \sqrt{-7}, -j6, -j4, j\pi$
2 $j7$ 3 j 4 $j6$
5 $j28$ 6 -12 7 $-j$
8 6 9 $j10$ 10 $-j19$

Test yourself 14.2

1 $6 + j5$ 2 $5 - j4$ 3 $-j7$
4 $7 + j$ 5 $a + 2b - j(b + 8)$
6 $-2\theta - j\pi$ 7 $5 + j5$ 8 $14 + j2$
9 $27 - j5$ 10 $-27 + j36$ 11 $0.7 + j0.1$
12 $1 - j0.5$ 13 $0.8 + j0.6$ 14 $0.7 - j0.4$

Test yourself 14.3

1 $x = 1 + j3, x = 1 - j3$
2 $x = 2.5 + j1.5, x = 2.5 - j1.5$
3 $5.39 \underline{/21.80°}$ 4 $7.62 \underline{/293.20°}$
5 $3.61 \underline{/123.69°}$ 6 $6.40 \underline{/231.34°}$
7 $4 \underline{/0°}$ 8 $4.24 \underline{/45°}$
9 $7 \underline{/90°}$ 10 $6.08 \underline{/189.46°}$
11 $4 + j3$ 12 $12 + j$
13 $-5 - j5$ 14 $3 - j2$
15 $-2 - j$ 16 $-3 + j7$

Test yourself 14.4

1 $2.82 \underline{/90°}$ 2 $4.84 \underline{/48.07°}$
3 $7.25 \underline{/0°}$ 4 $2.31 \underline{/236.08°}$
5 $12 \underline{/50°}$ 6 $35 \underline{/240°}$
7 $6 \underline{/5°}$ 8 $70 \underline{/-125°}$
9 $3 \underline{/30°}$ 10 $7.5 \underline{/-40°}$
11 $4 \underline{/35°}$ 12 $8 \underline{/250°}$

13 $16\underline{/60°}$

14 $81\underline{/400°} = 81\underline{/40°}$

15 $4\underline{/50°}$, $4\underline{/170°}$ and $4\underline{/290°}$

16 $6\underline{/50°}$, $6\underline{/140°}$, $6\underline{/230°}$ and $6\underline{/320°}$

17 $7e^{j2}$

18 $3.2e^{j22.5}$

19 $4^{j\pi}$

20 $2.7^{j3.93}$

Test yourself 14.5

1 $I = 10.37 + j0.85 = 10.40$ A$\underline{/4.69°}$

2 $Z_{IN} = 5.95\underline{/28.1°}$; $I = 84.1$A$\underline{/-28.1°}$. The circuit is not resonating, because the phase angle is not zero.

Chapter 15

Test yourself 15.1

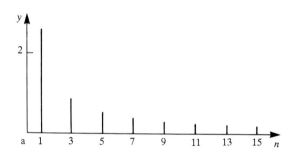

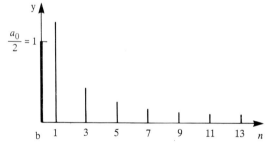

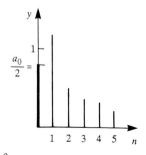

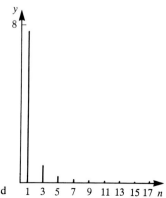

ANSWERS 311

1 a $y = -2, \quad 0 < t < \pi$
$\,y = 2, \quad \pi < t < 2\pi$

$a_0 = 0$, no initial (dc) term
$a_0 = 0$, no cosine terms
$b_n = -8/n\pi$, if n odd, odd sine terms present
$b_n = 0$ if n even, no odd sine terms

$$y = \frac{-8}{\pi}\left\{\sin t + \frac{\sin 3t}{3} + \frac{\sin 5t}{5} + \frac{\sin 7t}{7} + \frac{\sin 9t}{9} + \frac{\sin 11t}{11} + \frac{\sin 13t}{13} + \frac{\sin 15t}{15} + \ldots\right\}$$

b $y = 2, \quad 0 < t < \pi/2$
$\,y = 0, \quad \pi/2 < t < 3\pi/2$
$\,y = 2, \quad 3\pi/2 < t < 2\pi$

$a_0 = 2$, first term is 1
$a_n = 4/n\pi$, if $n = 1, 5, 9, 13, \ldots$
$a_n = -4/n\pi$, if $n = 3, 7, 11, 15, \ldots$
Odd cosine terms present, signs alternately $+, -, \ldots$
$a_n = 0$, if n even, no even cosine terms
$b_n = 0$, no sine terms

$$y = 1 + \frac{4}{\pi}\left\{\cos t - \frac{\cos 3t}{3} + \frac{\cos 5t}{5} - \frac{\cos 7t}{7} + \frac{\cos 9t}{9} - \frac{\cos 11t}{11} + \frac{\cos 13t}{13} - \ldots\right\}$$

c $y = t, \quad 0 < t < \pi$
$\,y = 0, \quad \pi < t < 2\pi$

$a_0 = \pi/2$, first term is $\pi/4$
$a_n = -2/\pi n^2$, if n odd
$a_n = 0$, if n even, no even cosine terms
$b_n = 1/n$, if n odd
$b_n = -1/n$, if n even
All sine terms present, with signs alternately $+, -, \ldots$

$$y = \frac{\pi}{4} - \frac{2}{\pi}\left\{\cos t + \frac{\cos 3t}{9} + \frac{\cos 5t}{25} + \ldots\right\} + \sin t - \frac{\sin 2t}{2} + \frac{\sin 3t}{3} - \frac{\sin 4t}{4} + \frac{\sin 5t}{5} + \ldots$$

d $y = 3\pi - 6t, \quad 0 < t < \pi$
  $y = -9 + 6t, \quad \pi < t < 2\pi$

$a_0 = 0$, no initial (dc) term
$a_n = 24/\pi n^2$ if n odd
$a_n = 0$ if n even, no even cosine terms
$b_n = 0$, no sine terms

$$y = \frac{24}{\pi}\left\{\cos t + \frac{\cos 3t}{9} + \frac{\cos 5t}{25} + \frac{\cos 7t}{49} + \frac{\cos 9t}{81} + \frac{\cos 11t}{121} + \frac{\cos 13t}{169} \right.$$
$$\left. + \frac{\cos 15t}{225} + \frac{\cos 17t}{289} + \ldots\right\}$$

Test yourself 15.2

2 a Half-wave repetition; no odd terms.
  b Half-wave inversion, odd function; $a_0 = 0$, no cosine terms, no even terms.
  c No symmetry as illustrated but, if $a_0/2$ is subtracted, it can be seen to have half-wave inversion and to be an odd function; no cosine terms, no even terms.
  d Even function; no sine terms.
  e Half-wave inversion; $a_0 = 0$, no even terms.
  f Half-wave inversion, even function; $a_0 = 0$, no sine terms, no even terms.
  g No symmetry as illustrated but if a_0 is subtracted it has half-wave inversion; no even terms.
  h Odd function; $a_0 = 0$, no cosine terms.
3 A sawtooth wave. Even function with half-wave inversion when $a_0/2$ is subtracted; no sine terms, no even terms.

$$y = \frac{\pi}{2} + \frac{4}{\pi}\left\{\cos t + \frac{\cos 3t}{9} + \frac{\cos 5t}{25} + \ldots\right\}$$

4 Square wave. Even function with half-wave inversion when $a_0/2$ is subtracted; no sine terms, no even terms.

$$y = 1.5 + \frac{6}{\pi}\left\{\cos t - \frac{\cos 3t}{3} + \frac{\cos 5t}{5} - \frac{\cos 7t}{7} + \ldots\right\}$$

Index

References to definitions and detailed explanations are given in **bold** type

Abscissa, **50**
Absolute values, **ix**, 147
Adjoint matrix, **233**
Amplitude, **65–6**, 269
Angle, **29**, 264
Angular velocity, **67**, **95**, 269
Antidifferentiation, **141**
Antilogarithm, **27**, 83
Antinode, **268–9**
Arbitrary constant, **175**
Argand diagram, **248–9**
Argument, **72**
Asymptote, **59**
Augmented matrix, **236**
Auxiliary equation, **171**
Averages, 152–4
Axis, **50**, 244

Balancing equations, 34–6, 75–7
Base number, **25**
Boundary condition, **159**, 161, 181

Cancelling, **13**
Cartesian coordinate, **50**
Chain rule, **127–9**
Circle, **69**
Circular motion, **95–6**, 97
Coefficient, **9**, 278–9
Co-factor, **217**, 233
Co-factor matrix, **233**
Column matrix, **225–6**
Commutative property, **230**
Complex frequency domain, **196**
Complex number, 173, **244**, 245–9
Complex number plane, **244**
Conjugate complex number, **246–7**
Constant,
   arbitrary, **175**
   of integration, **141–2**, 161
Convergence, **195**
Convergent series, **255**
Coordinate, **50**, 93
Cosine, **x**, 65, 67, 122, 124, 274
Counting number, **3**
Critical damping, 179
Cubic curve, **69**, 71

D (operator), **185**, 185–94
Damping, **176**, **177**, 179
dc component, 275, 280–1
Decay equation, **160–2**, 189

Decimal fraction, **4**, 46
Definite integral, 141, **144**, 274
Degree,
   of angle, **29**
   of expression, **18**
De Moivre's theorem, **254**
Denominator, **13**
Dependent variable, **49–50**
Derivative,
   first, **117**, 118, **122**, 126, 138
   second, 119, **124**, 126
Derived function, **117**, 138
Determinant, **213**, 231
Differential equation, **158**, 196–9, 204
   decay, **160–2**
   first order, 158, **163**, **166**, 190
   growth, **160**, **162**
   second order, **170–1**, 171–7, **178**, 191, 204
Differential,
   first, 117
   operator, 54, 114–5
   partial, **132–4**
   second, 119, **124**
   total, **136**
Dimension of s, **196**
Dirichlet conditions, **272–3**
Discriminant, **40**, 171–8
Divergent series, **255**
Dividend, **12**, 13
Divisor, **13**
Domain, **72**, 103

e (exponential operator), **5**, 26
Element, **213**
Elimination method, **41–3**, 212–3
Engineering form, **47**, 76, 84
Equation,
   auxiliary, **171**
   conditional, **34**
   differential, **158**, **160–3**, **166**, **170–1**, 171–8
   identical, **34**
   quadratic, **39**, 81, 248
   roots of, **39**, 171–3, 191–3
   simultaneous, **41**, 71, 78, 212–20, 235
   subject of, **37**, 75–6, 80
Equivalent fraction, **14**
Euler's method, **180–4**
Even symmetry, **286**, 287
Exponent, **23**
Exponential curve, **59**, 61–5, 89–90

Exponential form of complex number, 255–8
Exponential operator, **5**, 26
  series, **257**

Factor, **5**
  highest common (HCF), **6–7**
  prime, 5
Factorial number, **256**
Factorizing, 5–11
Family of curves, **159**, 181
Final value theorem, **209**
Finishing point, 28
First principles, differentiating from, **115–6**, 119
Fourier coefficient, **278–9**, 282, 283–4
  series, **272**
  transform, 273–9, 283
Fraction,
  algebraic, **16**
  decimal, **4**, 13
  equivalent, **14**
  improper, **13**, 79
  mixed, **14**
  partial, **18–22**
  vulgar, **13**
Fractional powers, **24**
Frequency, **67**
Frequency spectrum, **279**, 284
Function, **72**, **111**, 138
Function,
  composite, **127**
  derived, **117**
  even, **286**
  exponential, **201–2**
  inner, **127**
  odd, **286**
  outer, **127**
  piecewise, **73**, 275, 287
  periodic, **73**
  ramp, **200–1**
  simple, **127**
  sine, **202**, 274
  unit step, **199–200**
Fundamental, **268–9**

Gradient, **51**, 52–4, 70, 111, 124
Graph,
  cubic, **69**, 71
  exponential, **59**, 61, 89–90
  hyperbola, **69**
  linear, **51**, 56, 71
  logarithmic, **69**
  parabolic, **56–58**, 71
  polar, **94**
  straight-line, **51**, 56
Graph paper, 62, 64
Growth equation, **59**, **160**, **162**, 201–2

Half-wave inversion, **285**, 287
Half-wave repetition, **286**, 287
Harmonic, **268**, 279
HCF, **6–7**, 13
Hyperbola, **69**, 108

i (imaginary operator), **240–1**, 242–3
Identity, **19**, **34**, 68
  matrix, **230–1**, 234
  operator, **230**
Imaginary,
  axis, **244**
  number, **5**, 40, **240–1**
  number line, **242–4**
  operator, **240**
Improper fraction, **13**, 79
Independent variable, **49–50**, 269
Index, **23–4**, 82, 84, 88, 242, 253–4
Inflection, point of, **70**
Initial value theorem, **208–9**
Integer, **3–4**
Integral,
  definite, 141, **144**
  improper, **196**
  indefinite, **143**
  standard, 143, 148
  transforming, 206–8
Integrand, **143**
Integrating factor, **164**
Integration, **141**
Integration,
  by parts, 150–2
  by substitution, 147–9
  constant of, **141–2**, 143–5, 161
  ratios, 147
Intersection, point of, **71**
Intercept, **51**
Inverse, **23–4**, 29, 199
Inverse matrix, **230–1**, 232–4
  notation, **24**
Irrational number, **4**

j (imaginary operator), **240–1**, 242–3, 244

Laplace transformation, **195**, **203**, 258
LCM, **7–8**, 15, 16, 77
Limit,
  of definite integral, 141
  of expression, **103–6**
Logarithmic transformation, 61, 63, 195
Logarithm, **25**, 61–5, 82, 144–5, 258
Lowest common multiple, **7–8**, 15, 16, 77

Matrix, **225**
  adjoint, **233**
  arithmetic, **226–7**
  augmented, **236**
  co-factor, **233**
  column, **225–6**
  identity, **230–31**, 234
  inverse, **230–1**, 232–4
  multiplication, **227–30**
  row, **225**
  transpose, **233**
  vector, **28**, **225–6**
Maximum, **70**, 124
Modelling, **158**, 158–70, 174–7
Minimum, **70**, 124
Minor, **217**
Mixed fraction, **14**

INDEX

Natural logarithm, 26
Natural number, 3
Node, 268–9
Number,
   complex, 173, **244**, 245–9
   counting, **3**
   factorial, **256**
   imaginary, **5**, 40, 240–1
   integer, **3–4**
   irrational, **4**
   natural, **3–4**
   prime, **5**
   rational, **4**
   real, **4**, 40, 240
Numberline, **3–4**, 240
Numerator, **13**

Octave, 268
Odd function, **286**, 287
Operator,
   D, **185**, 185–94
   i, j, **240–1**, 242–3
Order, **23**, 213
Ordinate, **50**
Origin, **50**, **93**
Over-damping, **176**

π, **5**, 277
Parabola, 56–58
Partial differential, **132–4**
Partial fraction, **18–22**
Peak-to-peak, **66**
Percentage changes, 134–5
Percentage harmonic, **279**
Phase, **66**, 98–9
Phase angle, **66**, 86, 88–9, 99, 259–62
Phasor, 85, 86, 88–9, **99**, 249, 259–62, 263, 264
Period, **73**
Periodic function, 65, **73**
Pi, **5**, 277
Point of inflection, **70**
   of intersection, **71**
Polar coordinate, **93**, 100, 101
Polar form of complex number, **249**, 250, 263–6
Pole, **93**
Polynomial, **8**
Prime,
   factor, **5–6**
   number, **5**
Product, 130, 164
Product rule, **130**

Quadrant, **68**
Quadratic,
   equation, **39**, 81
   expression, **9**
   formula, 40, 81
Quotient, **12–13**, 130

Radian, **29**
Ramp function, **200–1**
Range, **72**

Rational number, **4**
Real axis, **244**
Real number, **4**, 40
Real part, **244**
Reciprocal, **23**, 77
Rectangular,
   coordinate, **50**, 93, 100, 101
   form, **244**, 249, 250, 263–6
   hyperbola, **69**, 108
Recurring decimal, **4**
Repetition, half-wave, **286**, 287
Resultant vector, **30**, 85, 249, **252**
Rms, **155–6**, 262
Root mean square, **155–6**, 262
Roots of complex numbers, 254–5
Roots of equation, **39**, 171–3, 191–3
Rounding, **46**
Row equivalent, **236**
Row matrix, **225**

Scalar, 227
Scientific form, **47**
Separable variables, **159**
Sequence, **255**
Series, **255**
   convergent, **255**
   divergent, **255**
   exponential, **257**
Significant figure, **46**, 75–6, 84
Simultaneous equation, **41**, 71, 78, 212–20, 235
Sine, **x**, 65–6, 84, 122, 124, 202, 274
Sine wave, **97–8**, 154–5, 264, **269**
Sinusoidal voltage, **97–8**, 154–5, 264
Solution,
   general, **175**
   of differential equation, **159**
   particular, **175**, 181
Standard form,
   of complex number, **244**
   of real number, **47**
Stationary point, **70**, 124
Starting point, **28**
Subject of equation, **37**, 75–6, 80
Substitution method, **43–5**
Symmetry,
   even, 286
   half-wave, **285**
   odd, **286**

Tangent,
   of an angle, **x**, 54, 65, 68
   to a curve, 58, **112**, 181
Total differential, **136**
Transform,
   Fourier, 273–8
   Laplace, **195**, **203**
   logarithmic, 195
Transient, **167**, 168, 169
Trigonometrical ratios, **x**, 65, 68

Under-damping, **176**
Unit step function, **199–200**

Variable, 46
Variable, separable, **159**
Vector, **28**, 96, 225, 248–9
Vectors,
    addition of, 30–2, 85, 248–9, 263, 264
    components of, **30**, 86–7, 97
    matrix, **28**, 225
    resultant, **30**

Vulgar fraction, **13**

Waveform,
    pulse, 276, 281, 283, 285
    sawtooth, 274–5, 276, 279, 282
    sine, **269**, 271, 286
    square, 276, 285
Whole number, **3–4**